The Play's the Thing

The Play's the Thing

Mathematical Games for the Classroom and Beyond

Alan Lipp

ANTHEM PRESS
LONDON · NEW YORK · DELHI

Anthem Press
An imprint of Wimbledon Publishing Company
www.anthempress.com

This edition first published in UK and USA 2011
by ANTHEM PRESS
75-76 Blackfriars Road, London SE1 8HA, UK
or PO Box 9779, London SW19 7ZG, UK
and
244 Madison Ave. #116, New York, NY 10016, USA

British Library Cataloguing in Publication Data
A catalogue record for this book is available from the British Library.

Library of Congress Cataloging in Publication Data
Lipp, Alan.
The play's the thing : mathematical games for the classroom and beyond / Alan Lipp.
p. cm.
ISBN 978-0-85728-666-6 (papercover : alk. paper)
1. Mathematical recreations. 2. Mathematics–Study and teaching. I. Title.
QA95.L764 2011
510.71–dc22
2010051140

ISBN-13: 978 0 85728 666 6 (Pbk)
ISBN-10: 0 85728 666 8 (Pbk)

This title is also available as an eBook.

Dedication

To my brother Michael who first taught me to love mathematics, and to my father Sunny who taught me to love teaching.

The Acts

Acknowledgments

Work on this book would never have been begun without the suggestions and encouragement of Jim Henle. And it would be far less useful or interesting were it not for hundreds of my students who have played these games with me and helped me to see what worked and what did not.

Prologue
To the Teacher

The goal of this book is to help students sharpen their reasoning skills, while having fun. This is a book about games; the fun comes both from playing the games and from discovering their secrets.

Each chapter, called an Act, describes a game and is followed by a few exercises aimed at helping students discover a winning strategy. Sometimes variations of the game are suggested, but eventually, we hope students will suggest their own variations. When students begin to create games and discover the strategies needed to win them, they will be creating mathematics.

Each section is written as a small self-contained unit to be completed in one or two periods as in-class work or over longer periods as independent study. If all variations are explored, some sections could take several days. Most of the games are independent and can be played in any order. Some, however, are used as models and revisited in later sections. The chart below gives the interdependence of the sections.

There are several broad mathematical themes that are heard again and again throughout the book: generalization, conjecture, investigation, and isomorphism. We encourage students to generalize our games and extend them, often from one dimension to two. Students are urged to search for winning strategies. This involves careful analysis of many games gathering data, observing patterns, conjecturing strategies, and then testing their conjectures.

In several cases games with different boards and different rules are actually, in some sense, the same (for example, Nimble and Two Piles). We call these hidden games; advanced texts would call them isomorphic games. With guidance, students will begin to see how one game may be lurking inside another and how strategies developed for one game can be used to find winning strategies in an isomorphic game.

The most important goal of the book, however, is to discover that doing mathematics can be fun. It is important that students be given time to play the games so that they can discover the logic of the game for themselves. Telling them the winning strategies will undercut both their enjoyment and their learning.

The first 18 Acts of the book are followed by a section called Harder Stuff, which includes more complex variations and more difficult questions about several of the games. Following the Harder Stuff is the Answer Key. While most questions are answered, a few are not. In some cases the answers are not known, but the question is presented in the belief that the search for an answer is worthwhile.

I hope that you and your students enjoy the book. Let the play begin.

Alan Lipp
South Deerfield, Massachusetts

Act 0

To the Student

This is a book of games. Some of the games are easy and some are quite difficult, but all of the games are meant to be played and enjoyed. This is also a book of mathematics. Mathematics isn't always about numbers and calculation or x and y. Mathematics is about patterns: finding them, describing them, and thinking about where they come from. A huge amount of mathematics has been written about games and the search for ways to win them because mathematicians enjoy both playing and understanding games.

In this book you will be asked to think about games and how to win them. Are some moves better than others? Why? Which ones? How can you tell? We hope that if you approach these questions playfully, you will learn a lot about the mathematics of games and have a lot of fun as well.

It's time to start Act 1. Enjoy the play!

Act 1
Blockers

We start with a simple game. Blockers is played on a long row of squares with two or more tokens placed anywhere on the board. Players take turns moving one of the tokens one or more squares to the left as many squares as they wish. They can also move a token off the board.

The tokens don't belong to the players so either player can move either token, however, no token may pass another token. That means that tokens near the left edge of the board block tokens further to the right. In Figure 1.1, for example, the token on square 6 can move only to squares 4 or 5; it cannot pass the token on square 3 or land on it. The winner is the last person who can move.

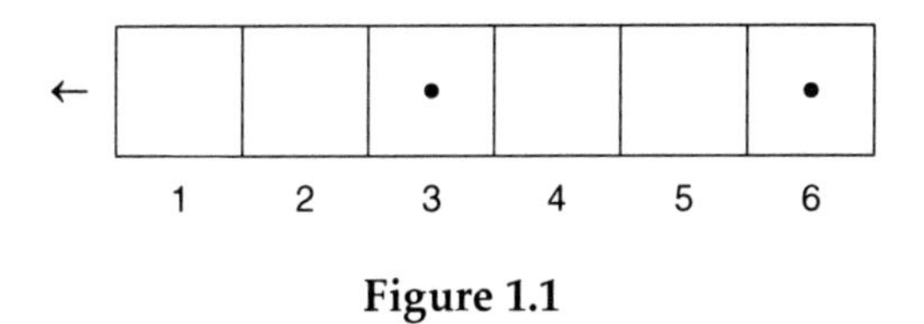

Figure 1.1

Find a partner and play Blockers several times on different length boards.

In most of the games in this book, a tie is not possible, so one player will always win. Of course, it is best to be that winner if you can!

Once you can figure out a pattern of moves that will guarantee a win, you have found a winning strategy. If it is your play and if you can find a winning strategy then you are going to win no matter what your opponent does. In this case, we call the position of the board a winning position because whoever plays next can force a win with good play. The Blockers game in Figure 1.1, for example, shows a winning position.

If a position is not a winning position, it is a losing position. If it is your turn to play in a losing position then no matter what you do, your opponent will win if she makes the best possible moves. In a game without the possibility of a draw, every position is either a winning position or a losing position.

Act 1 Exercises

1. The two players in any of the games in this book will usually be called A (the first player) and B (the second). Figure 1.2 shows a winning position. What is player A's winning move?

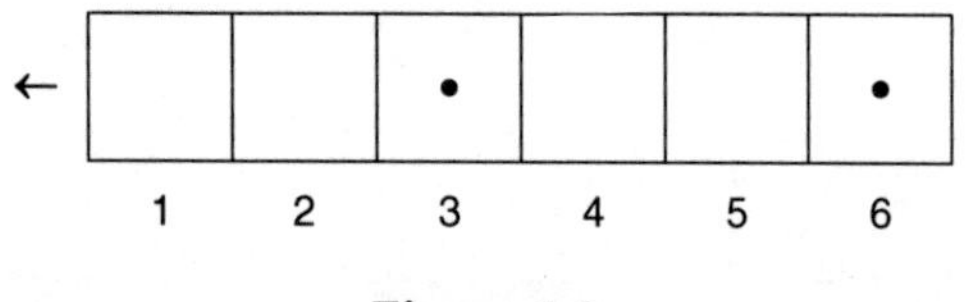

Figure 1.2

2. Blockers gets more complicated when the board gets longer. Is the position in Figure 1.3 a winning position or a losing position?

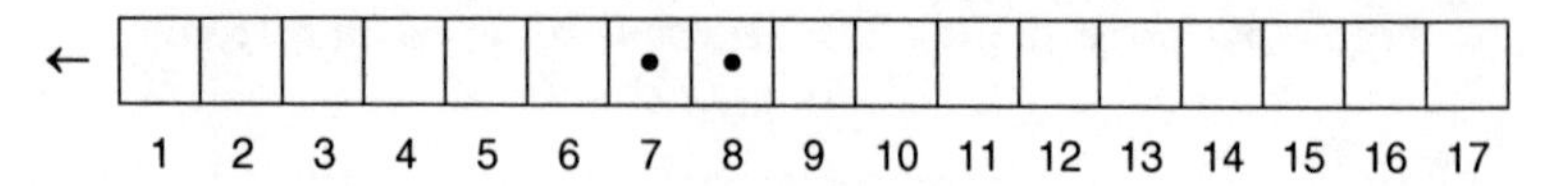

Figure 1.3

3. A winning move is only helpful if it is matched with a winning strategy. A winning strategy isn't a single move, it is an idea of how to respond to any possible move your opponent makes. After all, after A plays, B gets a chance to play and A doesn't know what B will do. Describe A's strategy so that A wins no matter what B does. Can you find a winning strategy for player A in Figure 1.4?

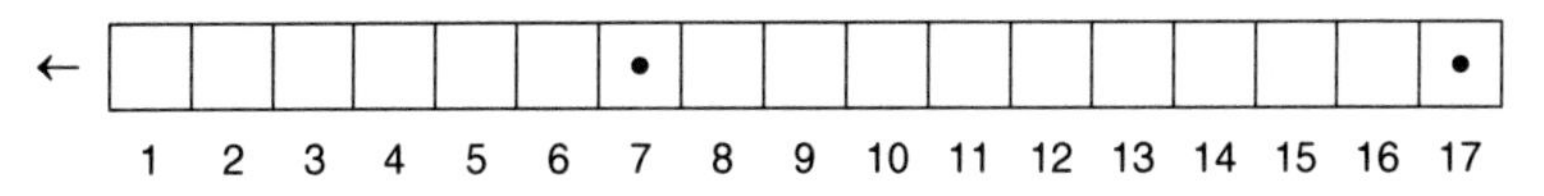

Figure 1.4

4. What is the winning move for A if the tokens start on squares 17 and 40?

5. Suppose the tokens start on squares n and $n + 5$ where $n \geq 1$. Is this a winning position or a losing position? Explain.

6. Describe the winning positions, and a winning strategy for 2-token Blockers.

Act 2
Nimble

Nimble is a game from the book *Fair Game: How to Play Impartial Combinatorial Games* (COMAP, 1989) by Richard Guy. It is a variation of Blockers in which we eliminate the blocking rule so that a player can move any token as far to the left as she wishes, including on top of another token or off the board. As usual, the first person that cannot make a move loses.

Figure 2.1 shows a position near the end of a Nimble game. The next player to move will lose if she moves either piece off the board because that would leave a move for her opponent. The best move, therefore, is to move the token on square 2 onto square 1.

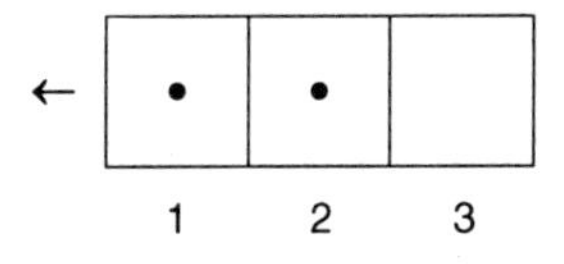

Figure 2.1

Find a partner and play Nimble several times on different length boards.

Act 2 Exercises

1. 6-Square Nimble is a little more complicated. Find the winning move in Figure 2.2.

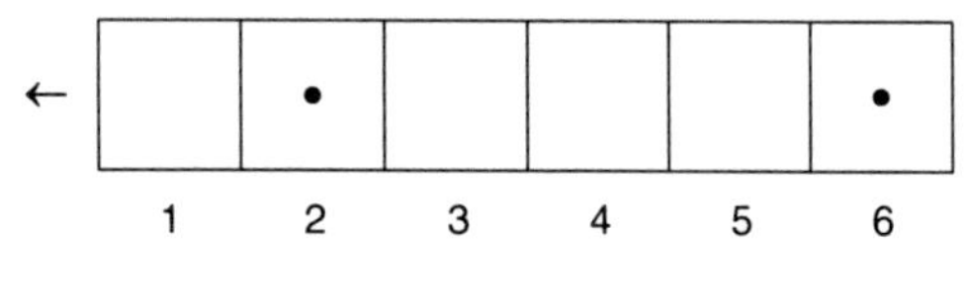

Figure 2.2

2. What is the best move in Figure 2.3?

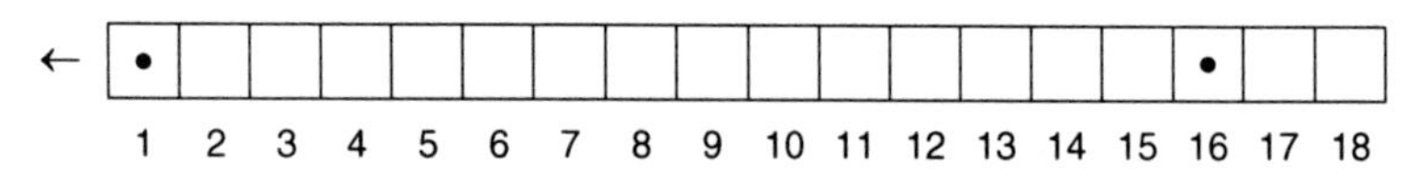

Figure 2.3

3. Figure 2.4 shows another board 18 squares long. Now what is the winning move?

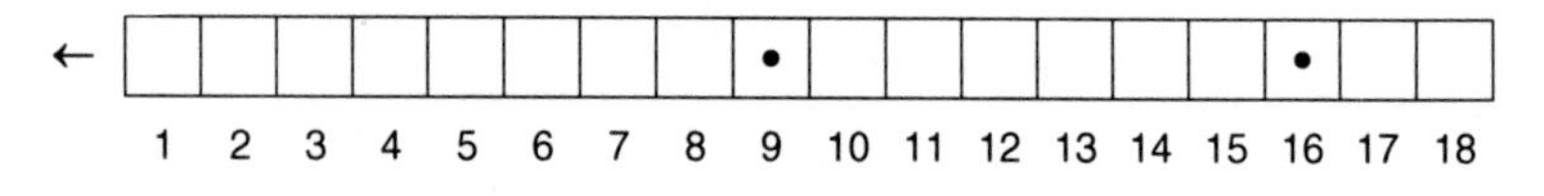

Figure 2.4

Have you figured out the winning strategy for Nimble yet? If you have not then you probably need to play a few more games before trying the rest of the challenges in this section.

4. Figure 2.5 shows another game board where one token is already in place. The second token can be placed on any of the 18 positions on the board. There is only one place to put the second token to form a losing position. Which square is that?

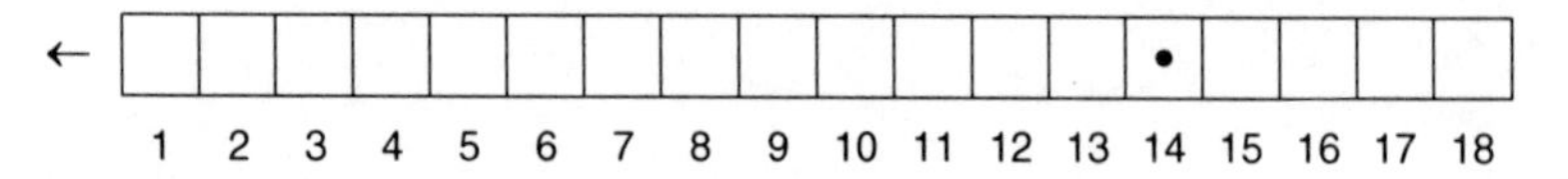

Figure 2.5

5. Nimble can be played on boards of any length. Describe the losing positions and the winning strategy in two-token Nimble that will work on any board.

So far we have only asked questions about strategy—about when and how we can win—but a game can provide many opportunities to pose other kinds of questions. Here are some variations that ask different kinds of questions. Perhaps you can think of other interesting questions.

6. The Longest Game.

   a) If you play Nimble as a solitaire game, so that you make all moves yourself, what is the longest possible Nimble game you could play on the 16-square board in Figure 2.6? Or, what is the largest total number of moves possible before the game is over?

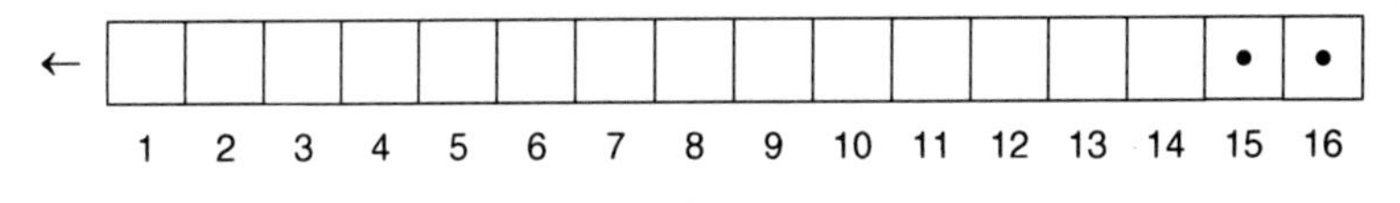

Figure 2.6

   b) Suppose the two pieces begin on squares m and n. Find a formula giving the largest number of moves possible in this game.

7. The Shortest Game.

   a) If you again play Nimble as solitaire, what is the smallest number of moves possible in the game in Figure 2.7?

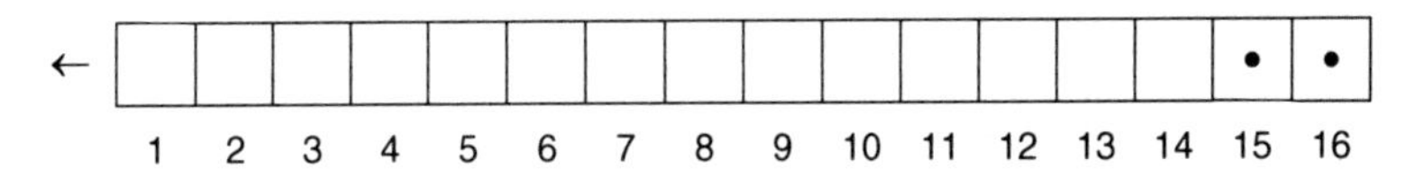

Figure 2.7

   b) If one piece is on square m and the other is on square n, what is the smallest number of moves possible?

Act 3

More Variations

You now know that the losing positions in 2-token Nimble have both tokens on the same square. The winning strategy from these positions is to keep the two tokens together, so whatever player A does, player B makes the same move with the other token. We call this strategy "copycat," since player B copies player A's move.

In this Act, we examine several variations of Nimble. Some of these will allow more than two tokens on the board, as in Figure 3.1. For convenience, we will use different rows for the tokens, as in Figure 3.2, which shows the same game. Now the game is determined by listing the locations of the tokens. We will refer to this position as [9, 9, 16] because the tokens in rows 1 and 2 are on square 9 and the token in row 3 is in square 16. If player A moves the token in the top row off the board, the new position is written as [–, 9, 16]. The dash indicates an empty row.

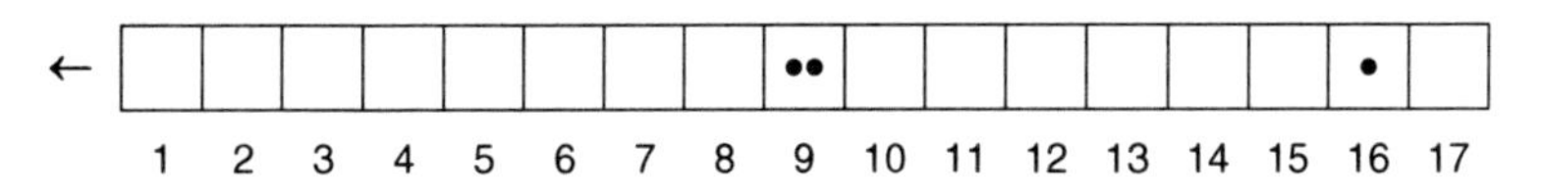

Figure 3.1

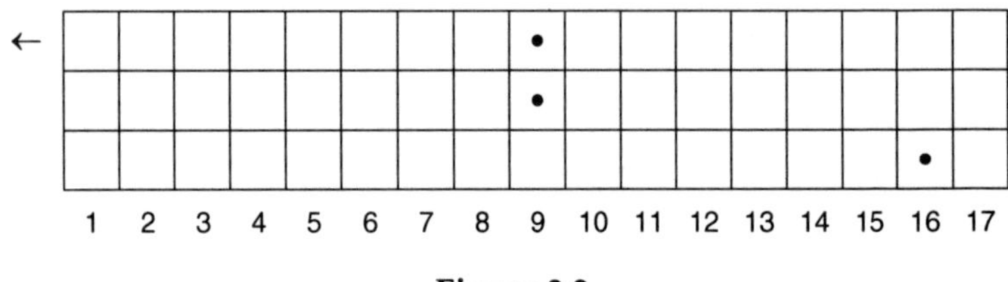

Figure 3.2

Be warned, the copycat strategy you discovered in Act 2 will need to be altered a little, so play with care!

Act 3 Exercises

1. The game board in Figure 3.3 shows position [9, 9, 16] in 3-token Nimble. If you play first, you can still be sure of a win. What is your winning move?

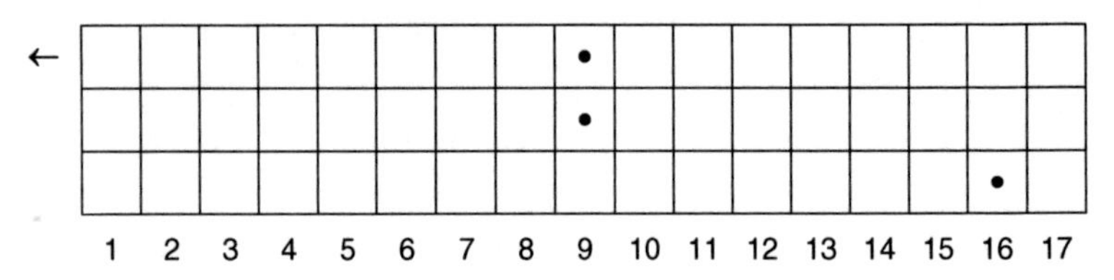

Figure 3.3

2. If you thought the winning move was to move the token on square 16 to square 9, making the position [9, 9, 9], think again. [9, 9, 9] is a *winning* position, not a losing position. Try again: what is the best move from position [9, 9, 16]?

3. Figure 3.4 shows [4, 10, 10], another winning position in 3-token Nimble. Find the winning strategy.

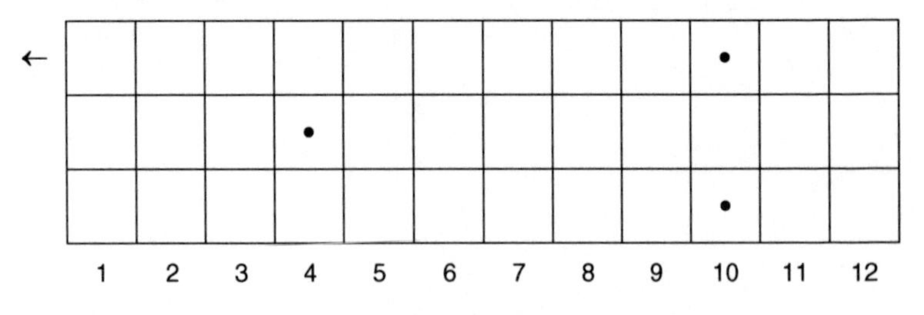

Figure 3.4

4. Is [–, 4, 10] a winning position or a losing position in 3-token Nimble?

5. Place 4 tokens on the board in Figure 3.5, 1 per row, so that player A has a winning strategy in 4-token Nimble. That is, create a winning position.

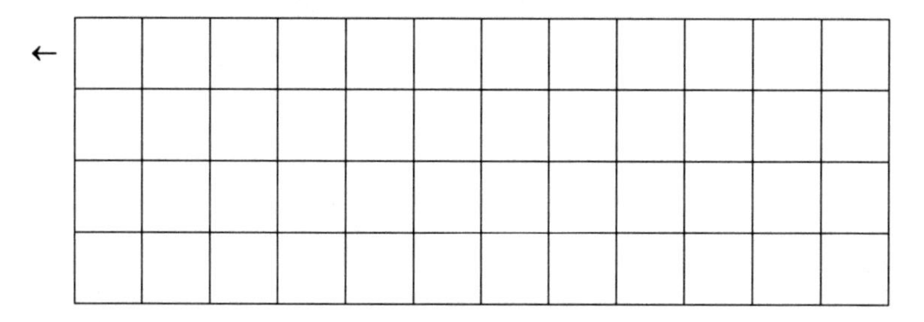

Figure 3.5

6. Place 4 tokens on the board in Figure 3.5 so that player B wins. That is, create a losing position.

7. $[n, n, m]$ indicates 2 tokens on square n and 1 token on square m in a 3-token Nimble game. If n and m are different numbers, is this a winning or a losing position?

8. Is $[n, n, n]$ a winning or a losing position?

9. Is $[n, n, n, n]$ a winning or a losing position?

10. Is $[n, n, n, n, n]$ a winning or a losing position?

11. Is a position with 22 tokens on square n a winning or losing position in 22-token Nimble?

12. Describe a winning strategy in $[n, n, m]$ Nimble.

13. Create your own Nimble variation. See if you can discover which positions have a winning strategy for the first player.

Act 4
Take Away 1, 2, 3

This game is played by taking away counters from a pile. Use pennies or pebbles for the counters, or whatever you have available. You can even play on paper, just by writing down the number of counters in the pile after each move. Moves alternate and players may take 1, 2 or 3 counters away from the pile. As usual, player A makes the first move. The first person that cannot play loses. Is there a winning strategy?

Find a partner and play 25 Take Away 1, 2, 3 several times. (That is, play with 25 counters.)

Looking for a Strategy

Take Away 1, 2, 3 is probably more complicated than Nimble and Blockers and it may not be obvious what the winning strategy is—or even if there is a winning strategy at all. When a problem is difficult, it is often a good idea to think about similar problems that are very simple. In this case, we might consider how to play Take Away in a game with only a few counters in the pile. This is a problem-solving technique that can sometimes help make a confusing problem simpler.

Suppose that there was only 1 counter in the pile—an extreme special case! This is obviously a winning position since the next player to move takes the counter and wins. If there were 2 counters in the pile, player A could lose if she foolishly took only 1 counter. But if player A makes the best move possible, she will take both counters and win. A pile with 1, 2, or 3 counters is a winning position because the next player can always win.

On the other hand, if there are 4 counters in the pile, whether player A takes 1 or 2 or 3 counters, she will be leaving a pile of either 3, 2 or 1 counters—a winning position for player B. A pile with 4 tokens is, therefore, a losing position.

Since Take Away 1, 2, 3 cannot end in a tie, every position is either a winning position or a losing position. Table 4.1 shows the winning and losing positions for a few small piles.

Pile Size	1	2	3	4	5	6	7	8	9
Win or Lose?	W	W	W	L	W	W	W	L	W

Table 4.1

Act 4 Exercises

1. Complete Table 4.2, showing the winning and losing positions for piles of up to 15. Describe the pattern you see.

Pile Size	1	2	3	4	5	6	7	8	9	10	11	12	13	14	15
Win or Lose?	W	W	W	L	W	W	W								

Table 4.2

2. Which size piles are losing positions?

3. Suppose we start with 175 counters. What is the best move? Why?

4. Knowing that a position is a losing position doesn't tell you how to win. What should player B's strategy be to win in 40-token Take Away 1, 2, 3?

5. Invent your own variation of Take Away.

Act 5

Two Piles: A Hidden Game

Scene 1

In Two Piles, we have 2 piles of tokens instead of 1. Players can remove any number of tokens from either pile, or even the entire pile. The first player who cannot move loses.

••••••••••••••••   •••••••

Figure 5.1

Figure 5.1 shows a game with 16 counters in one pile and 7 in the other. We will call this game [16, 7] Two Piles.

> **Find a partner and play Two Piles several times.**
> **Try playing with piles of various sizes.**

Scene 2

Now that you have played Two Piles, let's consider Figure 5.1 again. It isn't obvious what the best first move is. Does it matter which pile player A takes from? Does it matter how many counters she removes? It may help to organize the piles in rows to help us keep track of the size of the 2 piles. Now [16, 7] Two Piles looks like Figure 5.2.

•	•	•	•	•	•	•	•	•	•	•	•	•	•	•	•
•	•	•	•	•	•	•									
1	2	3	4	5	6	7	8	9	10	11	12	13	14	15	16

Figure 5.2

Does this figure remind you of anything?

"Two Piles" is an example of a *hidden game*. Taking tokens from any pile (row) is the same as moving the last token to the left. When the pile is empty, the token has moved off the board. Two Piles is Nimble in disguise! Here is [16, 7] Two Piles, drawn as a Nimble game.

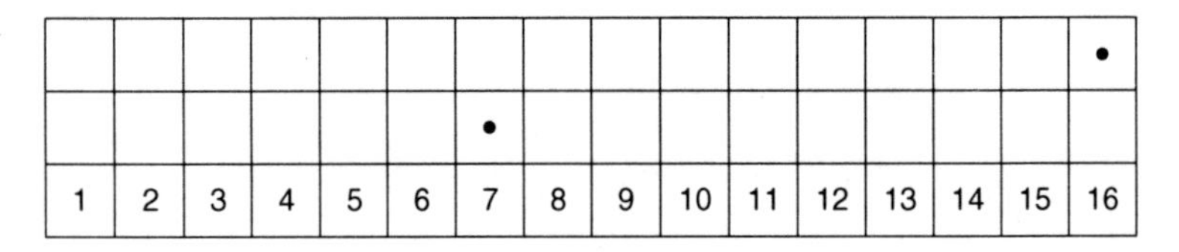

Figure 5.3

If you completed Act 1, then you should know the winning strategy in [16, 7] Nimble. Can you use it to design a winning strategy for [16, 7] Two Piles?

Scene 3

Mathematicians call games with such hidden similarities **isomorphic**. When you discover such hidden games, you can use the strategy from one game to find a similar winning strategy in the other. Since Two Piles is isomorphic to Nimble, we can modify Nimble's winning strategy to find a winning strategy in Two Piles.

The losing positions in Nimble are those with both tokens on the same square. The losing positions in Two Piles leave both piles equal. The winning Nimble strategy is to move the tokens to the same square. The winning Two Piles strategy is to make the piles equal.

Therefore, the winning move in [16, 7] Two Piles is to take 9 tokens from the larger pile. Whatever player B does, player A can win by playing copycat and matching the move that B makes. As long as A keeps the piles balanced, she will always have a move.

Scene 3 Exercises

1. What Two Piles game is isomorphic to this Nimble position?

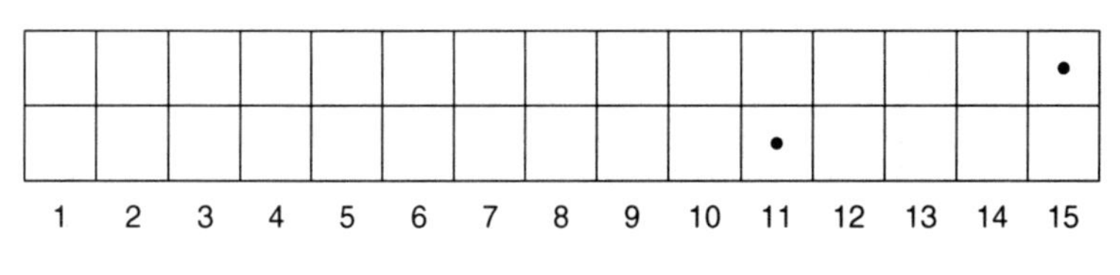

Figure 5.4

2. What is the best move in [5, 10] Two Piles?

Figure 5.5

3. Can you think of a new rule you could add to Two Piles so that it will contain Blockers as a hidden game?

Act 6

Two Piles 1, 2, 3

Scene 1

This game is similar to both Two Piles and to Take Away 1, 2, 3. We begin with 2 piles of tokens as in Two Piles but now players can only take 1, 2, or 3 tokens away from any one pile as in Take Away 1, 2, 3. We can vary the number of tokens in each pile from game to game.

Find a partner and play Two Piles 1, 2, 3 several times. Try playing with piles of various sizes.

Scene 2

Figure 6.1 shows a game with the starting position [5, 18], that is, with 5 tokens in one pile and 18 in the other.

Pile 1: • • • • •

Pile 2: • • • • • • • • • • • • • • • • • •

Figure 6.1

Does the first player or the second player have a winning strategy?

The second pile confuses the analysis; we cannot just count off in groups of fours as we did in Take Away 1, 2, 3 and we cannot just take away 10 tokens from the larger pile to balance as if we were playing Two Piles.

When a problem is confusing, it is often helpful to look at a simpler, similar problem. Keep the strategies of Two Piles and Take Away 1, 2, 3 in mind as you try these exercises.

Scene 2 Exercises

1. Is [18, 18] a winning position or a losing position? Why?
2. What is the winning move in [85, 87] Two Piles 1, 2, 3?
3. [24, 28] is a losing position. Can you explain why? What is player A's winning strategy?
4. What is the best move for player A from [88, 22]?

Scene 3

Equal piles are losing positions. From position [n, n], whatever player A does, player B can copycat to keep balanced piles. Whether A takes 1, 2 or 3 tokens, B takes the same amount from the other pile. This shows how the strategy of Two Piles can be used in Two Piles 1, 2, 3.

A different losing position occurs when both piles are multiples of 4. Whether A takes 1, 2 or 3 tokens, she will leave the untouched pile with a multiple of 4 tokens and the other with a number that is not a multiple of 4. Now B can imagine playing two different games of Take Away 1, 2, 3 at the same time. She should respond by taking from the same pile to make it into a multiple of 4 again. This is the "multiples of 4 strategy" of Take Away 1, 2, 3. Since this is a winning strategy, B will win in both Two Piles games.

Scene 3 Exercises

1. What is the best move in [5, 16] Take Away 1, 2, 3?
2. What is the best move from position [64, 47]?
3. What is the best move from the position $[n, n + 3]$?
4. What is the winning move from the position $[n, n - 2]$?
5. What is the winning move from position $[4n + 1, 8n]$?

Scene 4

We have now learned two different ways of thinking about Take Away 1, 2, 3. If one of the piles is a multiple of 4, then it helps to think of the game as if it were two simultaneous games of Take Away 1, 2, 3. The losing positions are those where each pile is a multiple of 4 and the winning strategy is to keep both piles as multiples of 4.

When the two piles are equal and not multiples of 4, however, a different strategy is required. Now we want to think of this as a game of Two Piles and try to keep the piles balanced and use the copycat strategy.

This brings us back to where we began in Figure 6.1 with starting position [5, 18]. These piles are not equal and cannot be made equal. They are not multiples of 4 and cannot both be made into multiples of 4. It is not obvious how to think about this game.

Scene 4 Exercises

1. Complete game table 6.1 for Two Piles 1, 2, 3 showing the winning and losing positions for pile sizes up to [12, 12]. The column headings are the number of tokens in Pile 1 and the row headings are the number of tokens in Pile 2.

	0	1	2	3	4	5	6	7	8	9	10	11	12
0													
1													
2													
3													
4													
5													
6													
7													
8													
9													
10													
11													
12													

Table 6.1

You can check your table by making sure that from each winning position you can move to a losing position by taking 1, 2, or 3 tokens from one pile and that from each losing position you can only reach winning positions. When you are sure that your table is correct go on to the next question.

2. Examine your table and try to describe the losing positions.

Scene 5

Did you discover the pattern of losing positions? In every case the *difference* of the piles is a multiple of 4. This includes 0, which is a multiple of 4. If you believe that this will happen, no matter what the sizes of the starting piles, then the winning move in any winning position is simply to make the difference in the piles a multiple of 4.

It is all well and good to discover a pattern, but it is even better to understand why that pattern works. When we understand a pattern we know for certain whether the pattern holds in all cases and, if not, what the exceptional cases are. Understanding the pattern in the completed game table is the goal of this scene.

Look at Figure 6.2, which again shows the starting position [5, 18].This time, however, we have rearranged the tokens.

(5) Pile 1: • • • • •

(18) Pile 2: • • • • • • • • • • • • • • • • • •

Figure 6.2

Notice that we separated the 18-pile into a 5-Pile and a 13-Pile. We can see that the two groups of 5 balance. The additional 13 tokens in the larger pile help us to think about the difference in the two piles. Ignore for a moment the two groups of 5 and think only of the extra sub-pile of 13. We can now use the strategies of both Take Away 1, 2, 3 and Two Piles to find winning moves.

Suppose player A takes 1, 2, or 3 from the 5 pile. Player B should copycat to keep the difference in the piles at 13. If A continues to take from the smaller pile, she will eventually reduce it to 0. At that point, it would be B's move with a pile of 13. B could win by taking 1 token to leave A with a multiple of 4 tokens in the only pile. Therefore, taking anything from the smaller pile is a losing move for A.

Is there a winning move for A if she takes from the larger pile? Suppose player A takes 1 token leaving [5, 17] for B. We can think of this position as [5, 5 + 12]. If B takes from the smaller pile, she will lose because A can use the strategy we outlined above. If player B takes from the larger pile, she must leave either [5, 5 + 11], [5, 5 + 10] or [5, 5 + 9]. In each case, the 5 pile is balanced and the difference in the piles is not a multiple of 4. Player A should take either 1, 2, or 3 tokens from the larger pile leaving [5, 13] which she can think of as [5, 5 + 8]. If B takes from the smaller pile, she loses because A can balance by playing copycat. If B takes from the larger pile, A can win by moving to [5, 9] or [5, 5 + 4], which keeps the difference as a multiple of 4. In this case, A is thinking about the winning strategy in Take Away 1, 2, 3.

Now we can see why piles that differ by multiples of 4 are losing positions. Whatever player A does, player B can think of the difference pile as a separate Take Away 1, 2, 3 game. If A took from the larger pile, then the difference is no longer a multiple of 4 and B can win by using the 'multiples of 4' strategy. If A took from the smaller pile, then player B can play copycat to balance the smaller pile. This leaves A with equal smaller piles (a losing position in Take Away) and with a difference pile that is a multiple of 4 (a losing position in Take Away 1, 2, 3). Whatever A does, B can win by playing the appropriate strategy.

Scene 5 Exercises

1. a) What is the winning move in [22, 57] Two Piles 1, 2, 3?

   b) If player B responds by taking 3 from the smaller pile, what should player A do?

   c) If instead, player B responds by taking 3 from the larger pile, what should player A do?

2. a) What is a winning move in [7, 22] Two Piles 1, 2, 3?

   b) If player B responds by taking 2 away from the smaller pile, what should player A do?

   c) If instead player B responds by taking 3 away from the larger pile, what should player A do?

3. Is [25, 50] a winning position? If so, what is the best move?

4. Would you prefer to play first or second in [20, 32] Two Piles 1, 2, 3? Why?

5. Can you think of a new rule you could add to Nimble so that it will contain Two Piles 1, 2, 3 as a hidden game?

Act 7
Nim

Scene 1

Two Piles is a special version of the game, Nim. Nim is played with any number of piles, each with an arbitrary number of counters. A Nim move consists of removing one or more counters from any pile. As usual, the last person to move wins.

Figure 7.1 shows a game of Nim with piles of size 5, 6, and 7, which we will shorten to [5, 6, 7] Nim.

Pile 1: • • • • •

Pile 2: • • • • • •

Pile 3: • • • • • • •

Figure 7.1

Find a partner and play [5, 6, 7] Nim several times.

Although [5, 6, 7] is a winning position, it is not easy to find the winning move. The winning strategy is much more subtle than the strategies we have seen so far and it will take some careful investigation to discover it. We have seen how useful the "balance and play copycat" strategy has been, but it isn't obvious how to balance an odd number of piles.

Scene 1 Exercises

1. Suppose player A tries to use the copycat strategy in [7, 6, 6] Nim (Figure 7.2) and so takes 1 token from the first pile to balance, leaving [6, 6, 6]. Show that player B has a winning strategy.

••••••• •••••• ••••••

Figure 7.2

2. Player A did not have to lose the game in Figure 7.2, since [7, 6, 6] is a winning position. Can you find the winning move?
3. [55, 48, 55] is a winning position. What is A's best move?
4. Show that $[n, n, m]$ is a winning position.

Scene 2

Player A is in error in exercise 1 in thinking that [6, 6, 6] is a losing position. If A makes the mistake of taking 1 from the 7 pile in Figure 7.2, she leaves [6, 6, 6], a winning position. Player B's best move is to take all of 1 pile, leaving 2 equal piles. Once 1 pile is gone, we can think of the result as a game of Two Piles, and [6, 6] is a losing 2-pile position. There is a winning move from [7, 6, 6] but taking 1 from the 7-pile is not it. Player A should have removed all of the 7-pile, leaving 2 balanced piles for player B—a losing position in Two Piles.

The idea of balancing is a powerful one. The problem is how to decide what "balanced" means when there are 3 piles. A Harvard mathematician named C. L. Bouton did just that about a hundred years ago. He realized that thinking about whole piles was not good enough. That idea will work for a position like [7, 7, 10] but not for [7, 8, 10], both of which are winning positions. It turns out to be helpful to think of each pile as made up of smaller piles the way we did in Take Away 1, 2, 3. Using Bouton's strategy, you can find a different winning move for player A in Figure 7.2. We'll save that question for the exercises.

Bouton got the idea of thinking about each pile as made up piles of size 1, 2, 4, 8, 16, 32, ..., each double the size of the pile before it. These numbers are all powers of 2: 2^0, 2^1, 2^2, 2^3… We'll call these smaller piles "binary sub-piles" because "binary" means two-number. Bouton's strategy has three steps.

1. Separate each pile into binary sub-piles using the largest size piles possible. For example, you would break 5 into 4 and 1, not into 2, 2 and 1. If you are familiar with the binary number system, this is the same as writing each pile in binary notation. Figure 7.3 shows how to break [5, 6, 7] into binary sub-piles.

Pile 1:	5	•••• •
Pile 2:	6	•••• ••
Pile 3:	7	•••• •• •

Figure 7.3

2. Bouton decided that it was these sub-piles that should be balanced—that is, paired off. So now we count the total number of piles of each size. In this case, there are three 4-piles, two 2-piles and two 1-piles. Keep track of which groups appear an odd number of times, since the others are already balanced.

3. The winning strategy is to balance *all* of the binary sub-piles. Do this by taking counters from *the largest pile containing an unbalanced sub-pile.* In this case, the largest pile containing a 4 is pile 3 and so the winning move is to take 4 from pile 3 leaving (5, 6, 3).

Scene 2 Exercises

1. We know now that taking all of the 7-pile will win in [7, 6, 6] Nim. Can you find a different winning move for player A?

2. It isn't obvious that [5, 6, 3] is a losing position, but it is. Suppose that player B takes away 1 from the 6-Pile. What should player A do to balance? Suppose B takes 2 away from the 3-pile. What should player A do to balance?

3. According to Bouton, is [6, 7, 4] a winning or a losing position?

4. Play [6, 7, 3] Nim with yourself as a puzzle. Start by making binary sub-piles. When you are player A, use Bouton's strategy to help you keep the sub-piles balanced. When you are player B, do anything you wish. See who wins. If you find a position where you cannot use Bouton's strategy, make note of the problem.

Scene 3

Do you see how Bouton's strategy allows the winning player to play copycat? Once the sub-piles are all balanced, no matter what the first player does there will be a kind of copycat response. Or will there? Let's try several more examples.

[14, 30, 25, 16] Nim is more complex since we now have 4 piles. When we divide these piles into binary sub-piles, we get Figure 7.4.

	••••••••	••••	••	14
••••••••••••••••	••••••••	••••	••	30
••••••••••••••••	••••••••		•	25
••••••••••••••••				16

Figure 7.4

There are an odd number of 16s, 8s and 1s but no pile contains all 3 types of sub-piles so we cannot simply remove 16 + 8 + 1 = 25 tokens from the largest pile. The problem here is not with the idea of balancing, but with how to carry it out. Bouton discovered a clever way to even out the piles in cases like this. In the exercises below, you will discover Bouton's method.

Scene 3 Exercises

1. Taking 25 from the largest pile will lose in [14, 30, 25, 16] Nim. Figure 7.4 shows the sub-piles. Take 25 from the largest pile and break up the new position into binary sub-piles. Explain how Bouton's strategy turns this into a winning position for player B.

2. In the previous problem, you found that taking 25 from 30 does not balance the binary sub-piles, even though 25 is the correct sum of the piles you want to remove. Look again at Figure 7.4. This time, we will try to balance the sub-piles one step at a time.

   Step 1: The 16s are unbalanced, so take 16 away from 30. Imagine that you are holding these 16 counters in your hand.

   Step 2: The 8s are unbalanced so take 8 away from the 30 pile, too. Now you have 24 counters in your hand.

   Step 3: We still need to balance the 1-pile, but the 30-pile doesn't have a 1-pile to remove.

   Find another way to balance the remaining 1-pile (still working with the 30 pile.) If you're stuck, think about the 24 counters still in your hand.

   If you figured out question 2, congratulations! You can skip down to the next exercise. If you are having trouble, you can skip to the end of these exercises where we explain Bouton's strategy, then come back and continue with problem 3.

3. A winning move in [14, 30, 25, 16] is to take 23 from pile 30. There are 2 other winning first moves. What are they? (Instead of drawing the piles, you might find it easier to just show how the piles break up into sub-piles: [(8, 4, 2), (16, 8, 4, 2), (16, 8, 1), (16)]).

4. Is [7, 2, 1] a winning position or a losing Nim position? If you think it is a winning position, what is the winning move?

5. Use Bouton's strategy to find all the winning moves in [25, 19, 11, 8] Nim.

6. What should player A do in [37, 29, 20] Nim?

7. a) Find a Nimble position that corresponds to [7, 20, 18] Nim (Figure 7.5).

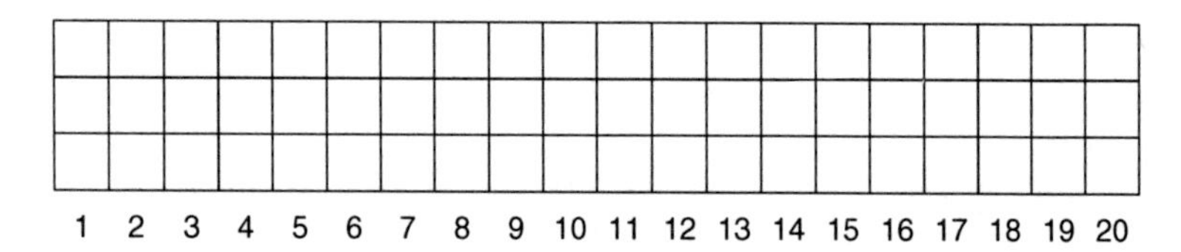

Figure 7.5

   b) Find a winning move for the first player in this position of Nimble.

   c) Is there another winning move?

8. Find 3 different winning moves if it is your move in the Nimble game in Figure 7.6.

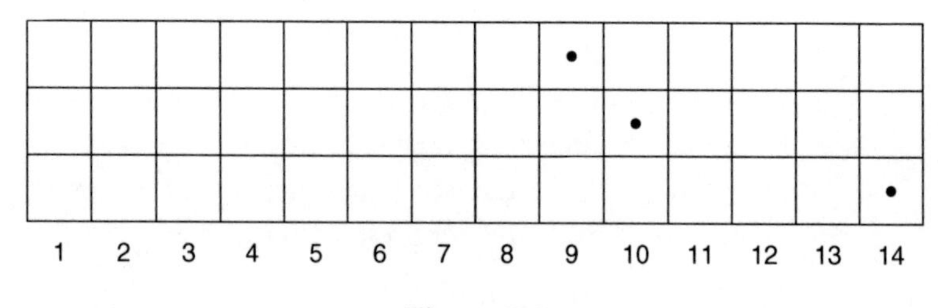

Figure 7.6

9. a) Place a token in the last row of Figure 7.7 so that the next player to play cannot win if her opponent uses Bouton's strategy.

b) Is there another place you could place a token in the third row to make a losing position?

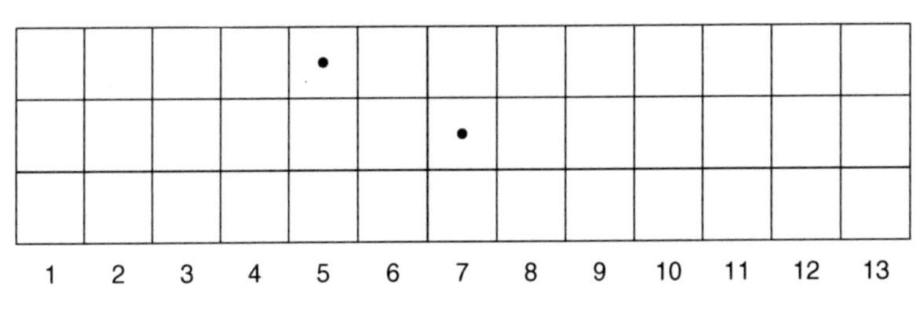

Figure 7.7

Bouton's Strategy

Bouton reasoned that he could balance an odd number of sub-piles, either by taking away a sub-pile or by creating a new sub-pile of the same size. The 4 piles [14, 30, 25, 16] break into [(8, 6, 2), (16, 8, 4, 2), (16, 8, 1), (16)]. We can balance the odd sub-piles by taking 23 from 30. We can think of this in 3 steps:

Step 1: Take away 16 from 30 to balance the 16 piles.

Step 2: Take away 8 from 30 to balance the 8-piles.

Step 3: Put back 1 from 30 to create an extra 1-pile to balance the 1-piles.

Bouton's Winning Strategy: Choose a pile that is large enough so that you can take away one of each of the odd-numbered groups that it contains, and *add back* to this pile each of the odd groups that it does not contain.

Act 8
Flit

"Flit" is short for "fill it." Flit is an example of a "placement game." Players take turns placing one or more pieces on a board, and the last person to be able to move wins. Differently shaped boards and different types of pieces create a variety of different games. We will begin by looking at One Row Flit, where the board is a single row of squares, as in Nimble.

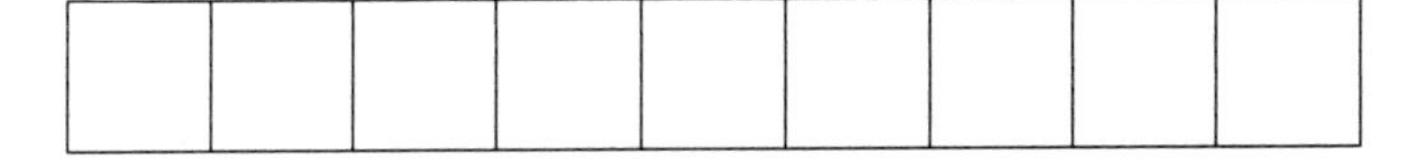

Figure 8.1

The simplest version is "One-At-A-Time" or "1Flit," where players alternate in placing a single token on the game board. Can you see why this game would be boring to play? The outcome of the game is completely determined by the number of squares on the board. If there are an odd number of squares, then the first player must win; if there is an even number of squares, the game belongs to the second player. There isn't even a strategy to the play, these wins happen automatically.

"Two-At-A-Time" or "2Flit," where players alternately place 2 touching pieces on the board, is much more interesting. On the above board, player A has seven choices for their first move.

Find a partner and play several games of 12-box 2Flit (That is, play Two-At-A-Time on a board with 12 squares).

The outcome of this game is not determined just by counting, since early moves may isolate several squares, eliminating them from play. For a simple example, suppose it is your turn to play on a 4 x 1 board. You have a choice of three possible moves (Figure 8.2). If you play to the left or to the right, you lose but if you play in the center, you win, since your opponent will not be able to play. This shows that the first player can always win 4-box 2Flit. Of course, the first player would also win 2-box and 3-box 2Flit as well.

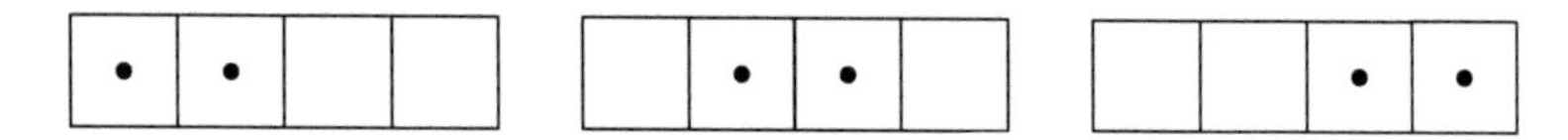

Figure 8.2

Act 8 Exercises

1. In Figure 8.3, which player will win 5-box 2Flit?

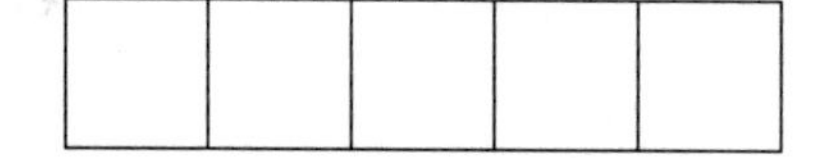

Figure 8.3

2. In Figure 8.4, does the first player have a winning strategy in 6-box 2Flit? If so, what is it?

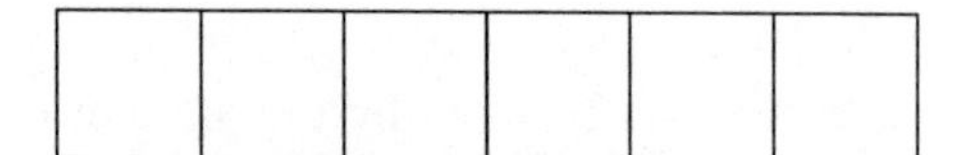

Figure 8.4

3. Extend the table below, showing the winning and losing positions in 2Flit for different size boards.

Number of boxes	2	3	4	5	6	7	8	9	10	11	12	13	14
Player A	W	W											

Table 8.1

4. Problem 3 suggests a general pattern of winning and losing positions. Who has a winning strategy in 2Flit with 30 boxes? 68 boxes? 81 boxes?

5. What is the winning strategy for even length boards?

6. Give a formula for the losing positions on boards of length n.

Act 9
Mr Flit

Scene 1

Flit gets even more complicated when there is more than one row on the board. In "Many Rows Flit" (or "Mr Flit"), pieces can now be placed horizontally or vertically. Figure 9.1 shows 2 different legal moves on a 2×3 board. For convenience, we've labeled the columns with letters and the rows with numbers so that we can refer to the squares on the board. The move on the left is in a1 and a2, whilst the move on the right is in b1 and c1.

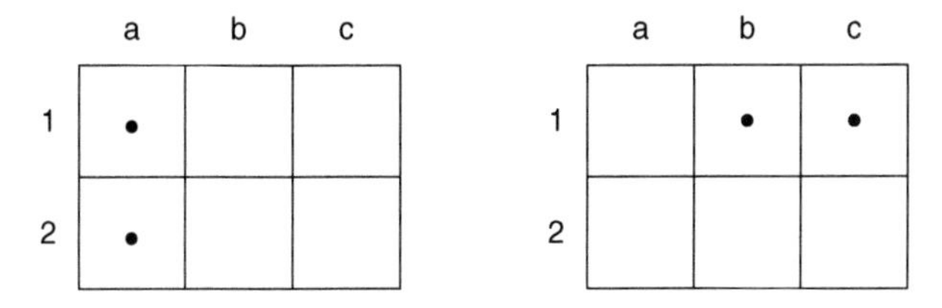

Figure 9.1

Find a partner and play several games of Mr Flit on a 5×1 board.

Scene 1 Exercises

1. Which player has a winning strategy on a 4×2 board (Figure 9.2)?

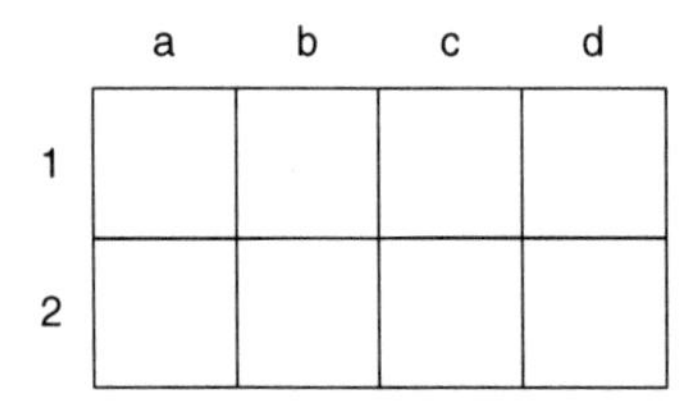

Figure 9.2

2. With the best play possible, who wins Mr Flit on a 3×3 board (Figure 9.3), player A or player B? Since we have not discovered a winning strategy for Mr Flit, you must try each possible sequence of moves to see who can win with best play.

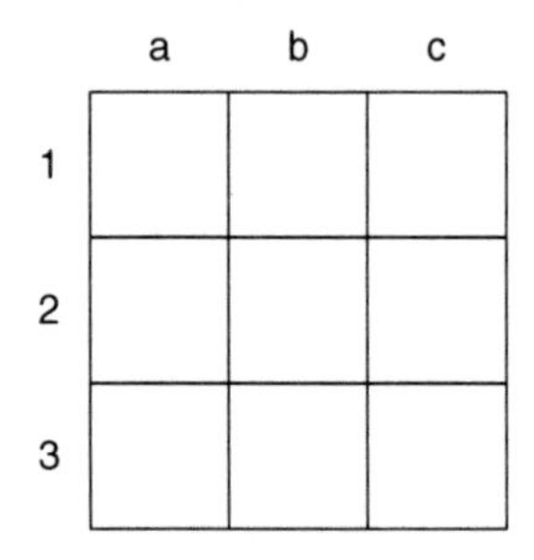

Figure 9.3

Scene 2

Mr Flit seems quite difficult on larger boards. In some cases, however, there is a simple winning strategy. Figure 9.4, for example, shows a 4×4 board. If player A moves in [a1, a2] player B can win. Can you find the winning strategy?

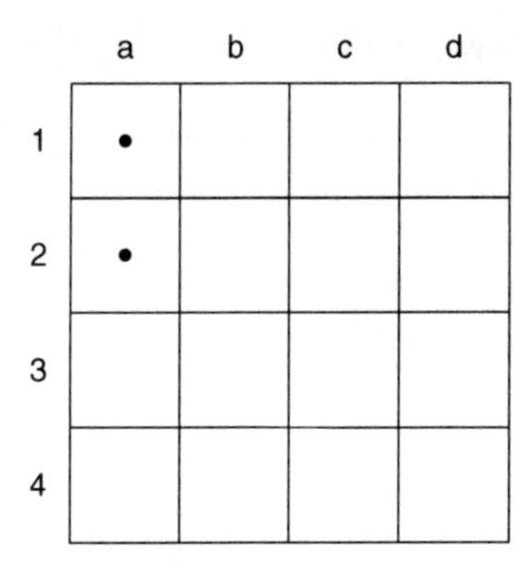

Figure 9.4

Remember that symmetry and copycat was the winning strategy for Flit. Do you see how that idea can be used here in Mr Flit? Notice that the board is symmetrical around its center point. Since A played in [a1, b1], player B can counter in [d3, d4]. Since every box has a symmetrical partner box, B can always follow A by playing copycat. Eventually, A will run out of moves and lose.

Scene 2 Exercises

1. Player A has a winning strategy on this 4×5 board. Can you find it?

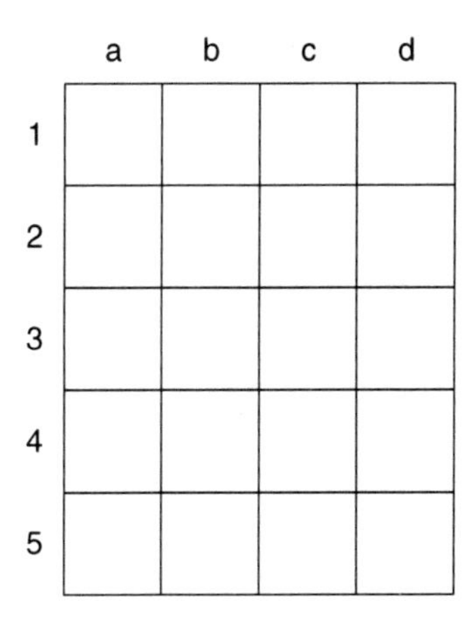

Figure 9.5

2. Which player has the advantage on a 24×13 board? Why?

3. Who has the advantage on a 12×20 board, player A or B? Why?

4. Who has the advantage on a board with an even number of rows and an odd number of columns? What is the winning strategy?

5. What has the advantage on a board with an even number of rows and an even number of columns? What is the winning strategy?

6. You may notice that in the last 4 questions, we did not ask about boards with an odd number of rows and columns. The reason is that these tend to be more complicated to analyze. Explain why that is.

7. **Variation 1: 1-2 Mr Flit.** Suppose players can play either 1 token or 2 adjacent tokens on the board.

    a) Which player has a winning strategy on an odd $\times$ odd board? That is, both an odd number of rows and an odd number of columns. What is the winning strategy?

    b) Which player has a winning strategy on an even $\times$ odd board? What is the winning strategy?

    c) Which player has a winning strategy on an even$\times$even board? What is the winning strategy?

8. **Variation 2: 2 Row Mr Flit.** Suppose we play Mr Flit on a $2 \times n$ rectangle but a move consists of placing 3 tokens in a row, not 2. Which player has a winning strategy?

9. **Variation 3: 3 Row Mr. Flit.** Suppose we play Mr Flit on a $3 \times n$ rectangle with a move placing 3 tokens in a row. Which player has a winning strategy if n is odd? If n is even, what is A's best first move?

10. **Variation 4: Ringers.** Suppose Mr Flit is played on a ring of squares, like the one in Figure 9.6.

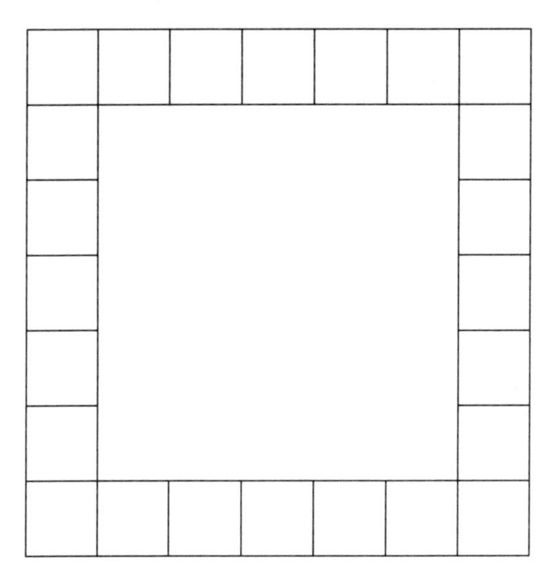

Figure 9.6

a) Which player has the advantage on this board?

b) This example has an odd number of squares on each side of the ring. Which player has a winning strategy on an odd×odd ring?

c) Which player has a winning strategy on an even×even ring?

d) Which player has a winning strategy on an even×odd ring?

e) Does the winning strategy change if players can choose to play either 1 or 2 tokens?

11. Create your own variation of Mr Flit and play it several times. Can you find a winning strategy for your game?

Act 10
Landis

"Landis" (or "L-and-Is") is a placement game like Mr Flit. Instead of both players using 2 tokens, in Landis, player A plays 3-in-a-row in an I-shape, whilst player B shades in 3 boxes in an L-shape. Figure 10.1 illustrates these shapes on a 7×5 board. Landis, like Mr Flit, can be played on any size board. As usual, the last person to be able to play is the winner.

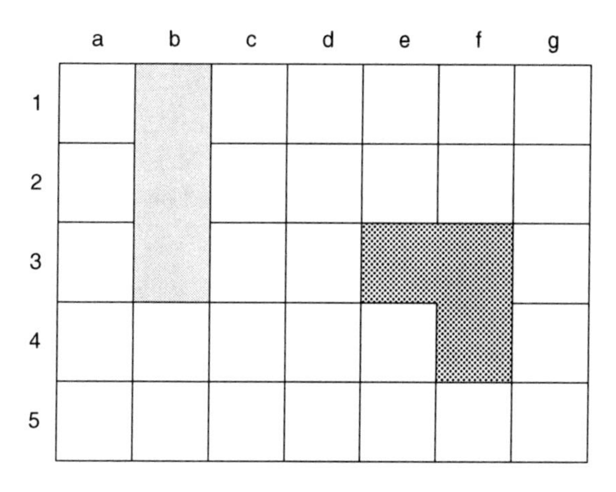

Figure 10.1

It is best to play the game several times before trying to figure out a general strategy, and when you begin to think about strategy it is usually easier to try simple boards first.

Find a partner and play several games of Landis. Graph paper will help.

Act 10 Exercises

1. In Figure 10.2, A played in the upper left corner and player B followed with their first piece. Where should A play in order to win?

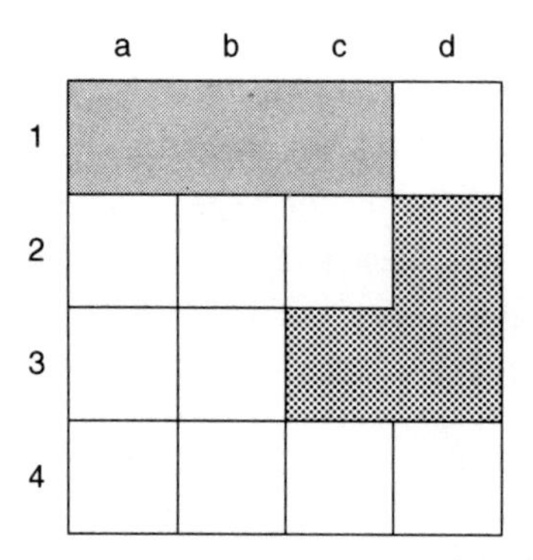

Figure 10.2

2. Figure 10.3 shows another first move. This one is also a poor choice. Why?

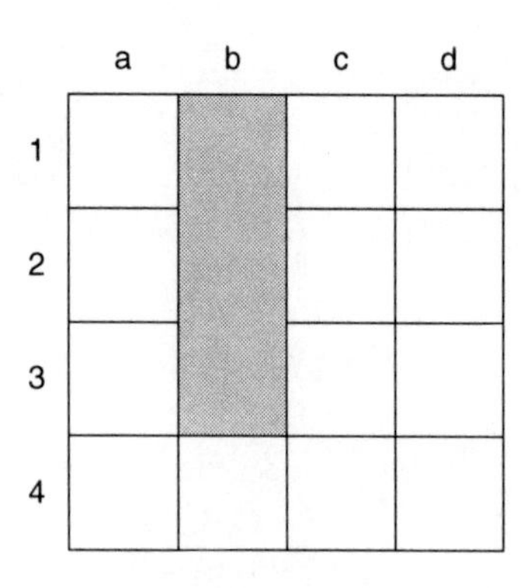

Figure 10.3

3. **The Longest Game**. The 5×5 game is more challenging because there are so many more possible places to play. If both players cooperate instead of trying to win, what is the most number of moves possible?

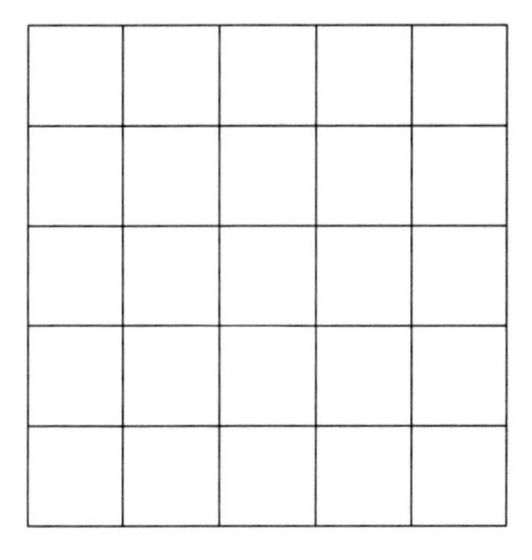

Figure 10.4

4. **The Shortest Game**. If both players cooperate, what is the fewest number of moves possible in 5×5 Landis?

5. It is I's turn to play in Figure 10.5. She can win the game provided she covers 1 particular square. Which square and why?

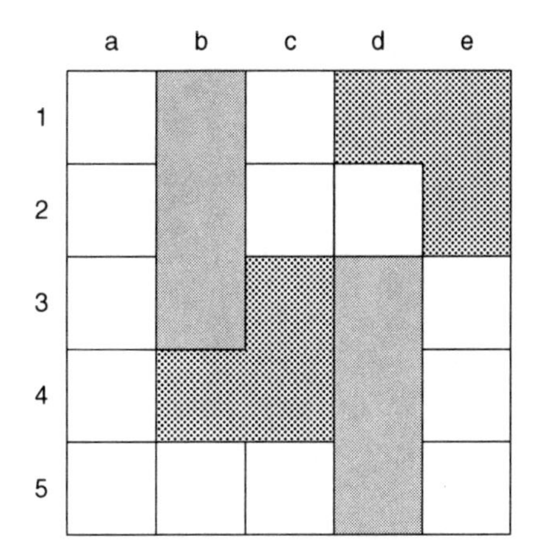

Figure 10.5

6. **Variation 1: 4×4 I-I Landis**

   a) Now both players place I pieces on a 4×4 square. I-I Landis is the same game as Mr Flit played with 3 tokens in a row or column. How many different possible first moves are there in a 4×4 board for player A to choose from?

   b) Does the first or the second player have a winning strategy? What is it?

7. **Variation 2: 4×4 L-L**

   a) Both players place L-shapes on a 4×4 board. Now, how many different first moves are possible?

   b) Why should A play around the center?

Act 11
Add'em Up

Add'em Up is a very different kind of game from the ones we have examined so far. Add'em Up is a played with numbers, not boxes. We begin by writing down a row of digits—for example, 8 4 5 6 3 8 9. A "move" consists of adding any 2 adjacent digits and replacing them with their sum. If the sum is a 2-digit number, the digits are written separately. In the above row, for example, we could add 8 + 9, transforming the board into 8 4 5 6 3 1 7. Turns alternate and the winner, as usual, is the last player who has 2 numbers to add.

For example, in the game 5 6 7, player A could either replace 5 6 with 1 1 or 6 7 with 1 3 (because 5+6=11 and 6+7=13). The game might continue as follows:

5	6	7	Player A adds 6 and 5.
1	1	7	Player B adds 1 and 7.
1	8		Player A adds 1 and 8.
9			Player B loses.

Since it is impossible for player 2 to move, player A is the winner.

Find a partner and play Add'em Up several times.
Try playing with different starting numbers.

Act 11 Exercises

1. There are five possible sequences that the game 8 8 8 can follow. Can you find all five sequences? What is the last digit left in each case?

2. Is 8 8 8 a winning position or a losing position for player A?

3. There are three possible sequences the game 4 6 8 can follow. What are the three possibilities, what is the last digit, and who wins?

4. a) Find a 3-digit game in which the first player loses.

   b) What property do all 3-digit games share, in which the first player loses?

5. How many moves are there in the game 1 1 1 1? In the game 1 1 1 1 1? In the game 1 1 1 1 1 1? Can you have a guess at the number of moves in a game with n1s?

6. **Total Variation.** As each player moves, they keep a running total of their sums. The game ends when no more additions can be made but the winner is now the player with the highest total. For example, the game 9 3 4 could continue 1 2 4 with player A scoring 12, or 9 7 with player A scoring 7.

   a) What is the final score of each of the 5 possible 8 8 8 games you found in question 2?

   b) Player A has a winning strategy for 8 6 7. What should be her first move be if she hopes to make the greatest score possible?

7. **Total Solitaire.** Play Total Variation as a solitaire puzzle. If you make all the moves and keep a running total of your sums, what is the greatest total you can reach starting with:

   a) 1 2 3 4 5 6 7 8 9

   b) 9 9 9 9 9 9 9 9 9

   c) 9 8 7 6 5 6 7 8 9

8. **Many Digits Add'em Up**. Suppose you are allowed to group the adjacent digits as you please—as single digits, 2-digit numbers, 3-digit numbers, or more. For example, the first move in 1 2 3 4 5 6 7 8 9 could be to replace 3 4 with 7, or to replace 3 4 5 6 with 9 0, thinking 34 + 56 = 90, or 1 2 3 4 with 127. Play this version several times.

   Does the first player always have a win in 1 2 3 4 5 6 7 8 9? We don't know the answer—perhaps you can discover it.

9. If you play Many Digits, scoring as you did in problem 6, what is the highest score you can achieve on the first move, starting with 8 8 7 7 6 6 5 5 7 7 8 8?

Act 12
Connect-the-Dots

This classic is played on dot paper with a rectangular array of dots. A move consists of joining any two unconnected adjacent dots. (Dots are adjacent if they are one box away, either vertically or horizontally. Dots cannot be connected diagonally.) If a player completes the fourth side of a 1×1 box, she gets a point and plays again. The game ends when all dots have been connected; the winner is the player with the most points.

Since it is time consuming to draw dot paper, you may prefer to use graph paper and use the intersections of the graph lines as the dots.

The rule allowing a player an extra turn whenever she completes a box is what makes this puzzle so interesting and so difficult to analyze. Turns do not strictly alternate as they do in every other game we have considered. The game has been analyzed for certain boards, but not in general.

Act 12 Exercises

1. The usual game is played on a 9×9 board. The game often lasts a while because there are so many lines to draw. How many segments must be drawn to completely fill a 9×9 board?

2. a) Connect-the-Dots is easier to analyze if the board is smaller. For example, the first player must lose on the smallest possible board—2×2 dots. Why?

   b) The next board is 2×3. There are 7 possible first moves. Does one of them guarantee a win for player A?

3. Suppose you are playing Connect-the-Dots on a 3×3 board. It is your move in Figure 12.1. What is your best move?

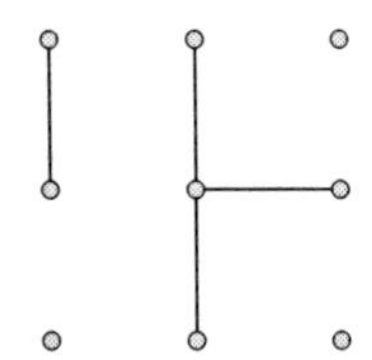

Figure 12.1

4. A 2×4 board (Figure 12.2) has only 3 squares to form so the game cannot end in a tie. With the best possible play, who must win?

•   •   •   •

•   •   •   •

Figure 12.2

5. **All Squares Count.** In this variation of Connect-the-Dots, players get a point whenever they complete a square *of any size*. If a line completes more than 1 square, the player scores for both. Now, a 3×3 board has **5** squares (4 squares with a side of 1 and 1 square with a side of 2), so the game cannot end in a tie. Who has the advantage—the first player or the second player?

•   •   •

•   •   •

•   •   •

Figure 12.3

6. **Long Lines.** In this variation, players can connect any 2 dots in the same row or column, providing their line does not overlap a shorter line in the same row or column. Players still score 1 point for each 1×1 box they form. In Figure 12.4, player A could join a1 to d1 and player 2 could respond with b4 to b1.

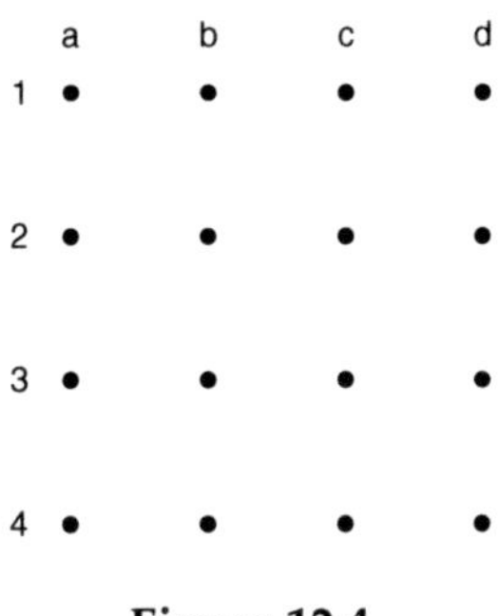

Figure 12.4

Play Long Lines several times on a 4×4 board. We do not know a winning strategy. Can you discover one?

If you are interested in learning more about Connect-the-Dots, you may enjoy the book, *The Dots-and-Boxes Game: Sophisticated Child's Play*, by Elwyn R. Berlekamp (A K Peters/CRC Press, 2000).

Act 13
Boxes

Boxes is another drawing game, but this time a player moves by drawing a box on a grid. Figure 13.1 shows a possible first move on a 2×6 game of Boxes, but the game can be played on any size rectangle. Players alternate drawing boxes. Boxes can be drawn inside or around other boxes and may touch another box at a corner, but they cannot intersect or share an edge with another box. We will name the boxes using the coordinates of the upper left and lower right corners. In Figure 13.1, player A has moved a2 to c3, but this is not the best first move, since player B can now win. Do you see how?

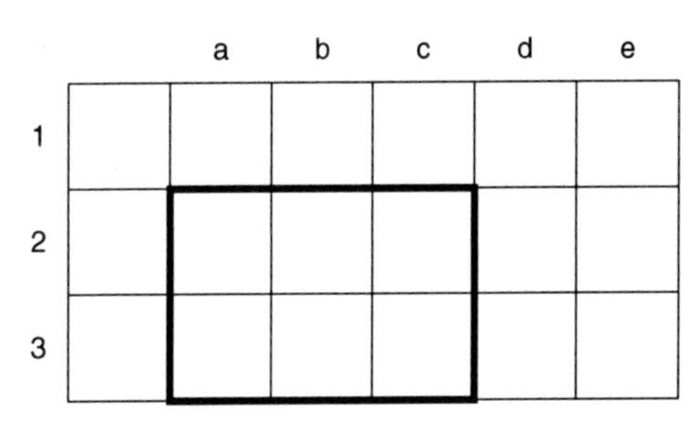

Figure 13.1

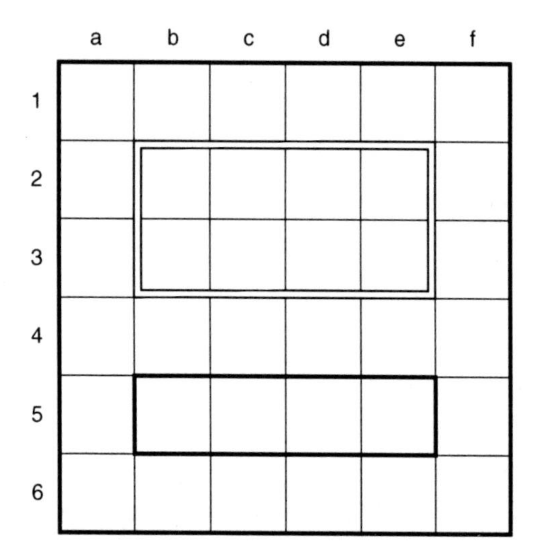

Figure 13.2

Figure 13.2 shows 3 moves on a 6×6 board. To help you to see the sequence of moves, player A's boxes are drawn with heavy lines, whilst player B's boxes are drawn with double lines. Player A began by drawing [a1, f6] around the whole board. Player B followed with [b2, e3]—a 2×4 box. A won the game with [b5, e5]—a 1×4 box—since it leaves no room for another box on the board.

> **Find a partner and play Boxes on a 10×10 board to get a feel for the game before trying the exercises. Graph paper will help a lot!**

Act 13 Exercises

1. Figure 13.3 shows a 1×11 board. What is player A's winning move on *any* $1 \times n$ board?

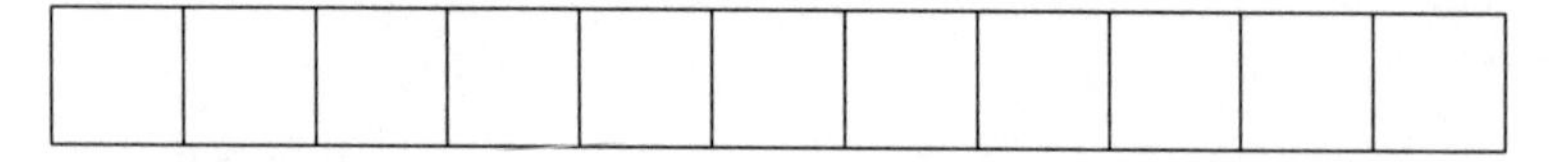

Figure 13.3

2. Figure 13.4 shows a 2×15 board. What is player A's winning move on *any* $2 \times n$ board?

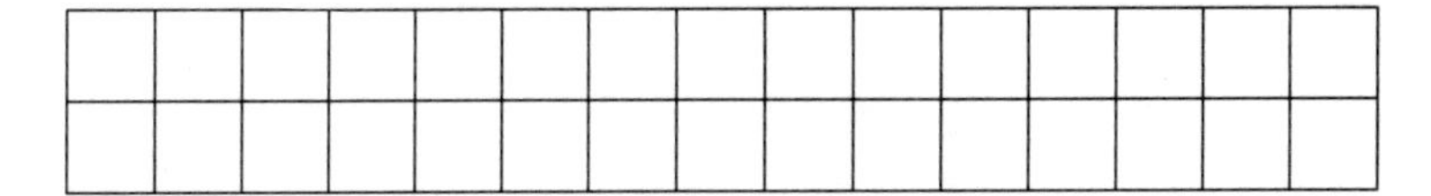

Figure 13.4

3. Figure 13.5 shows a 3×11 board. What is player A's winning move on *any* $3 \times n$ board?

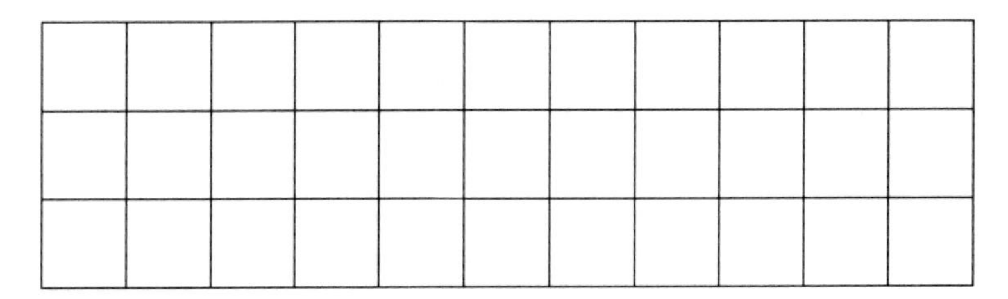

Figure 13.5

4. Figure 13.6 shows a winning position for player B. What is her winning move?

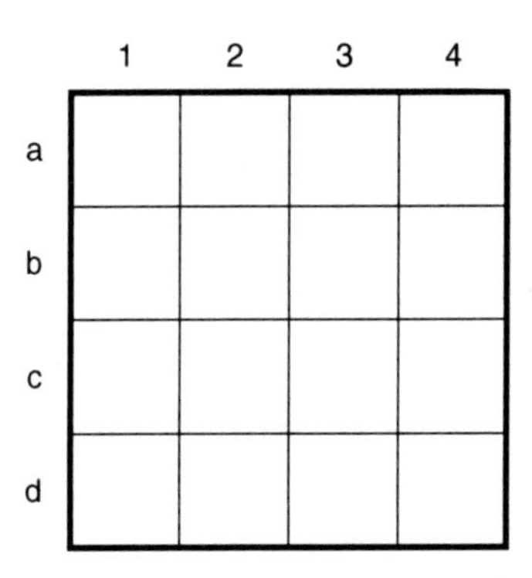

Figure 13.6

5. a) Figure 13.7 shows a winning position for player B. What is her winning move?

b) What (poor) move could player B make so that player A could win?

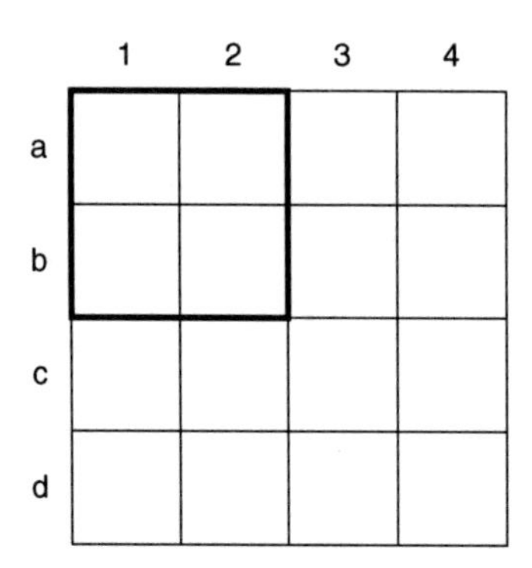

Figure 13.7

6. Figure 13.8 shows the first two moves in a 4×4 game of Boxes.

   a) Player A has a winning move. Can you find it?

   b) Is there any move that player A can make that will let player B win?

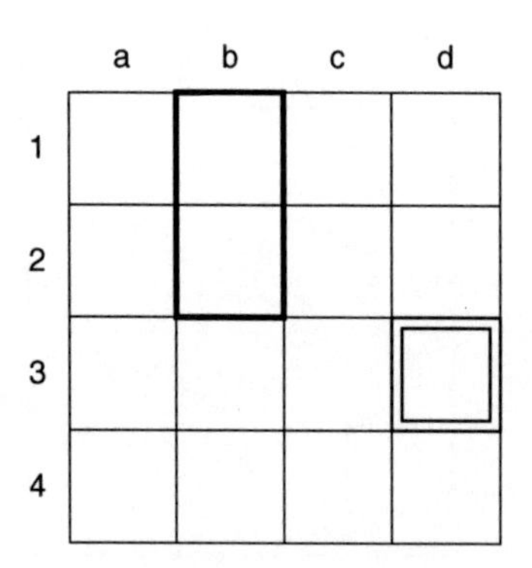

Figure 13.8

7. Figure 13.9 shows the first two moves in a 4×10 game of Boxes. What move could player A make that would allow player B to win?

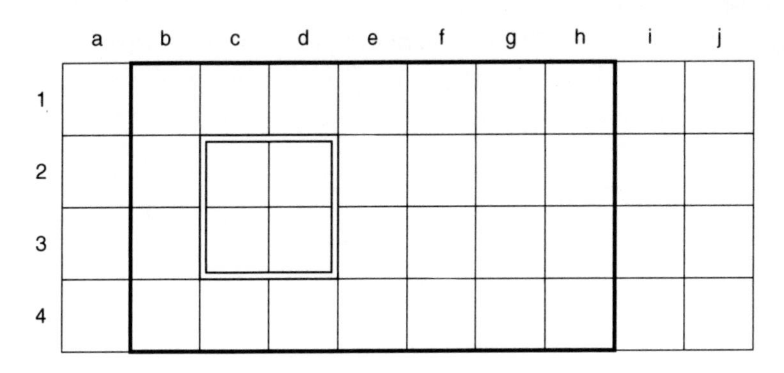

Figure 13.9

8. Player B can win in Figure 13.10. What is her winning move?

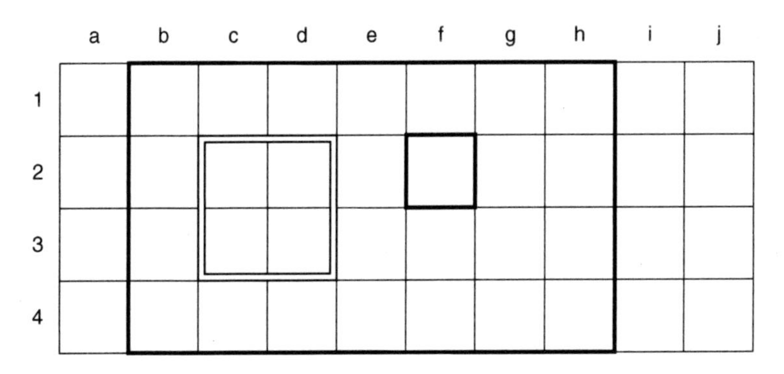

Figure 13.10

9. a) What is the greatest possible number of moves that can be played on a 10×10 board?

   b) What is the fewest number of moves possible on a 10×10 board?

10. a) What is the greatest number of moves possible on an even sided square board $2n \times 2n$? What is the greatest number of moves possible on an odd-sided square board $(2n-1) \times (2n-1)$?

b) What is the fewest number of moves possible on an even-sided square board $2n \times 2n$? What is the fewest number of moves possible on an odd-sided square board $(2n-1) \times (2n-1)$?

Can you generalize the questions in 9 to an $n \times n$ board?

Act 14
Hold That Line

The game "Hold That Line" was invented by Sid Sackson and published in 1969 in his book, *A Gamut of Games*. It can be played on any rectangle of points. The first player draws a line between any two points on the grid, even if the line passes over other grid dots. On the second turn, player B draws a line from one of the ends of player A's move to another point on the board. Next, A draws a line from one of the two endpoints to a new point. The only restriction is that new lines cannot cross over any other line on the board. In this way, the players alternately extend the broken line on the board until one of them cannot extend from either end. This is the losing player, since she cannot play.

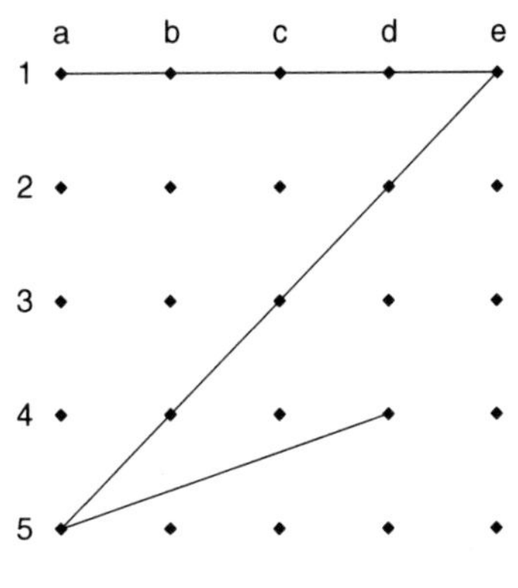

Figure 14.1

Figure 14.1 shows the first several moves in a typical game. As usual, we refer to the points on the grid by their column letter and row number. First, player A joined e1 to a5 and then player B joined b1 to e1. Player A then connected a5 to d4.

Player B could now join d4 to any of the points below the diagonal or b1 to any of the points above the diagonal.

Find a partner and play Hold That Line several times.

Act 14 Exercises

1. Does player A or B have a winning strategy in Hold That Line, played on a 2×2 board? How many moves will it take to win?

2. In 3×3 Hold That Line, how many possible first moves are there?

3. Suppose the board looks like Figure 14.2 after 2 moves in a 3×3 game. What move could player A make that would allow player B to win? How should player B continue in order to win?

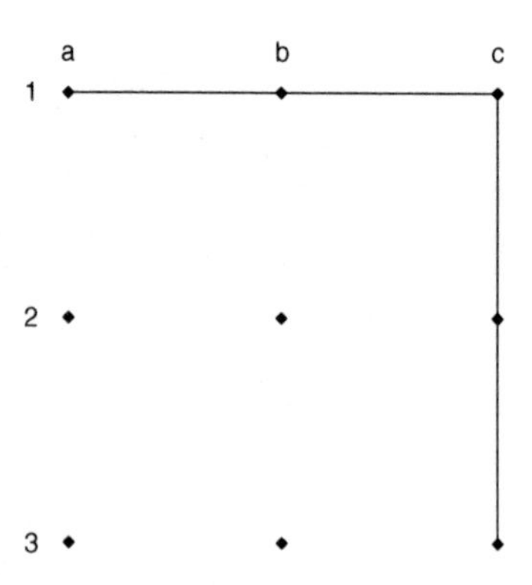

Figure 14.2

4. The next player has a winning move. What is it?

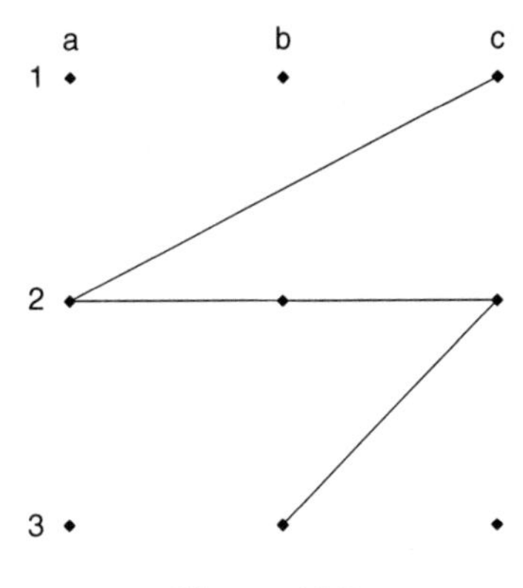

Figure 14.3

5. If both players cooperate, what is the greatest number of moves possible in 3×3 Hold That Line? What is the fewest number of moves possible?

6. Carefully describe a winning strategy for the first player in 3×3 Hold That Line.

7. Can you find a winning strategy for the first player in 4×4 Hold That Line?

8. Can you find a winning strategy for the first player that will work on *any* square board?

9. **Variation 1: Crossing Lines**. Suppose that we now allow lines to cross, so that a player can draw to any unused dot from one of the free ends of the lines already drawn. Can you find a winning strategy for one of the players?

10. **Variation 2: Isolated Lines**. Suppose that no line can touch any other line on the board. A move now consists of joining two unused points. Describe a winning strategy for player A on a 3×3 board.

Act 15

The Fifteen Game

The Fifteen Game holds a secret. It is played on a board with 9 squares, labeled 1 through 9. Players alternately place tokens on a number, one token per number. The winner is the first player who can make a total of 15 with any three of their numbers.

Figure 15.1 shows the starting board. We are using a heart shape (♥) for player A and a club shape (♣) for player B.

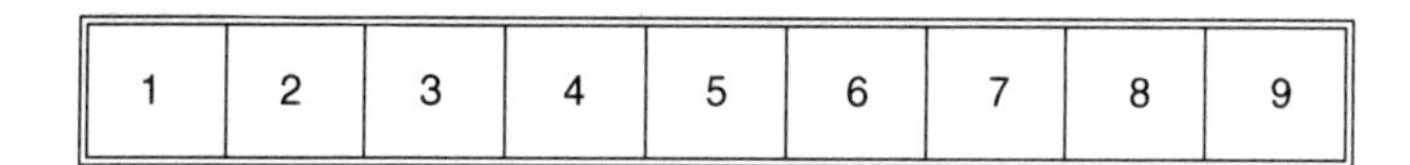

Figure 15.1

A typical game might play as follows. Player A takes the 4 and then B chooses 8 (Figure 15.2).

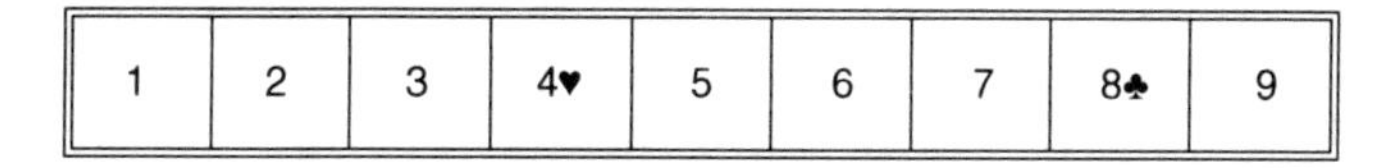

Figure 15.2

Now A plays on 2, giving her a total of 6. Since she needs 9 more to make 15, player B blocks by placing her token on the 9 (Figure 15.3).

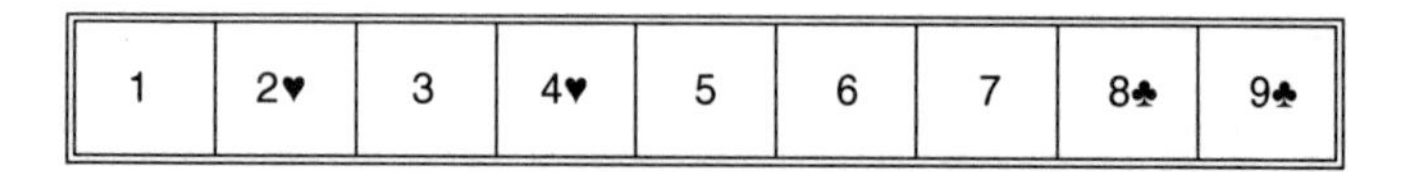

Figure 15.3

For her third move, player A takes the 6—a winning move (Figure 15.4). Can you see why?

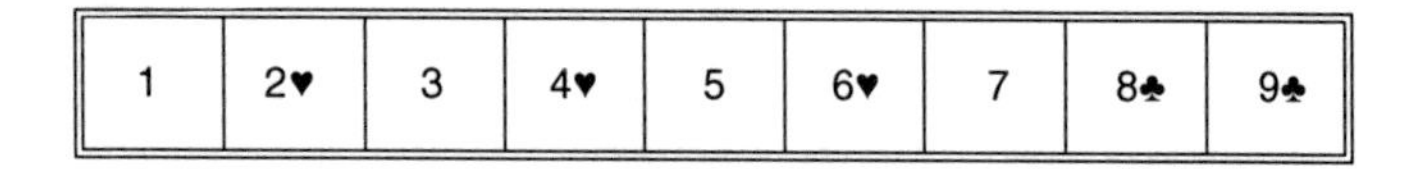

Figure 15.4

Player A is threatening to win in two different ways. Since $4 + 6 = 10$, she can win with a 5. But since $2 + 6 = 8$, she can win with a 7. Player B cannot block both threats. She will lose the game no matter what she plays now.

Find a partner and play the Fifteen Game several times.

Act 15 Exercises

1. Figure 15.5 shows a game after the first two moves.

1	2	3♣	4	5	6	7	8	9♥

Figure 15.5

   This is a winning position. What should player A do to win? Explain how this move guarantees the win.

2. A and B start a new game. Player A takes 4 and B takes 8. Player A takes 6 and claims it is a winning move. Is this correct? Why?

3. Player A again starts with 4 but this time B takes 2. Player A now has several different moves that will guarantee a win.

Can you find two of them? Explain what the winning strategy is in each case.

4. Player A starts with 5 and B moves onto 7. Show that player A has a winning strategy.

5. The Fifteen Game is the first game we have examined that can end in a tie. Find a sequence of moves that results in a game with no winner.

6. The Fifteen Game is another example of a hidden game. To see what game is hidden, look at 15.6, where the board has been rearranged into a square.

6	7	2
1	5	9
8	3	4

Figure 15.6

a) The numbers seem to have been placed in strange locations. What unusual mathematical property does this arrangement of numbers have?

b) What game is hidden in Figure 15.6?

c) Instead of hearts and clubs, what symbols are usually used for the two players in the hidden game?

7. Use Figure 15.6 to help answer this question. If A first plays on the 5 and then B responds with 3, where should A to force a win? Explain.

8. If player A begins with 9 and B plays 7, what is A's winning strategy?

Act 16
Sliders

Sliders can be played on any rectangular grid. Figure 16.1 shows the starting position on a 4×4 grid with tokens on every square.

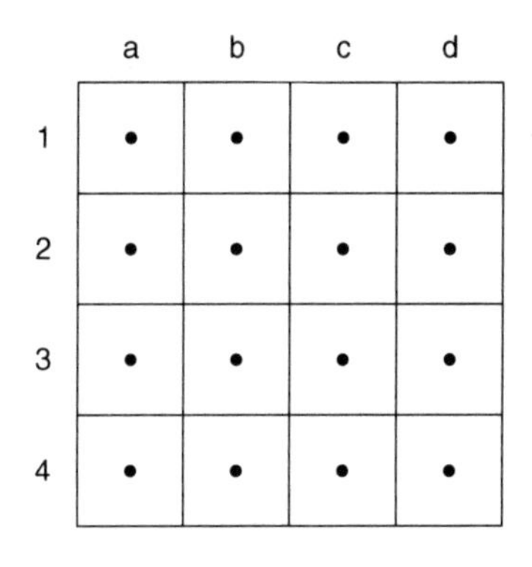

Figure 16.1

A move consists of sliding any token in any direction: up, down, left, or right. If there are empty spaces in front of the token, the move ends by leaving the token on one of the empty spaces or by sliding it off the board as in Nimble. If there are tokens in front of the mover then all of the adjacent tokens act as a train and slide together one or more spaces. Every slide must end either by stopping at a token or at the edge of the board or by dropping one or more tokens in the train off the board. Figure 16.2 illustrates these possibilities. Player A can move g2 to d2, e2, f2, h2 or to g1, or she could slide g2 off the top of the board. She could also move g2 downward to g3, causing the token at g4 to slide off the board. If player A moves a4 to the right, it can end on b4, c4, or d4, but she cannot push the token at c4 onto the token at g4.

	a	b	c	d	e	f	g	h	i	j	k	l
1	•	•	•	•								
2	•	•	•				•		•	•	•	
3							•					
4	•	•	•				•					

Figure 16.2

The two boards below illustrate the first two moves in a typical game of Sliders. For convenience, we've replaced the tokens with numbers:

1	2	3	4
5			6
9	10	11	12
13	14	15	16

1	2	3	4
5			6
12			
13	14	15	16

Figure 16.3

In the first move, player A pushed 6 to the right pushing 7 and 8 off the board. Player B responded by pushing 12 to the left and causing 9, 10, and 11 to move off the board.

Player A could now remove the entire top row or the entire left column by pushing all four tokens off the board. She could push 6 upwards, removing the 4 or removing both 4 and 6. If she pushes 6 downward, however, she can only move it one space, until it touches token 16. The last player to be able to move wins.

Find a partner and play 3 × 3 Sliders several times. When you are ready try the exercises below.

Act 16 Exercises

1. Who has the advantage in 2 × 2 Sliders (Figure 16.4)?

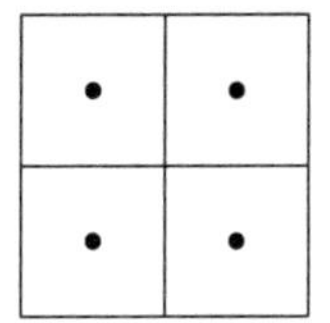

Figure 16.4

2. Suppose that player A removes row 1 of a game of 3 × 3 Sliders and player B responds by sliding row 2 from #4 to #6, leaving the L-shape shown in Figure 16.5. What is player A's winning strategy?

1	2	3
4	5	6
7	8	9

4	5	6
7	8	9

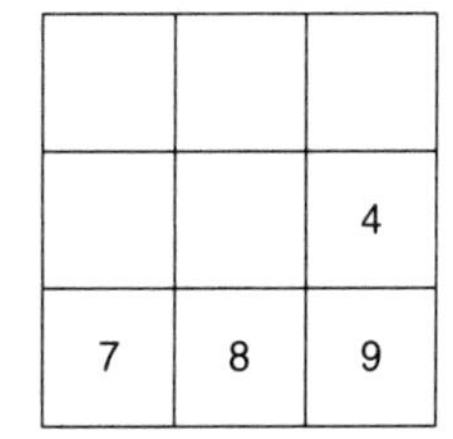

Figure 16.5

3. In problem 2, if player B moves 4 to 5 instead of 4 to 6, she leaves the position shown in Figure 16.6. Does player A still have a winning strategy? If so, what is it?

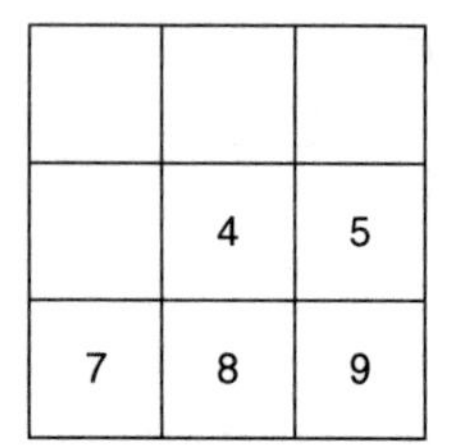

Figure 16.6

4. Does Figure 16.7 show a winning or a losing position in 3×3 Sliders?

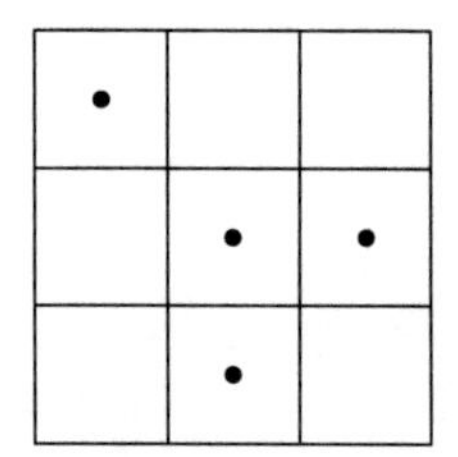

Figure 16.7

5. Will the symmetry strategy that works so well on other square boards work in 4×4 Sliders?

Act 17
Lynch

One way to make new games is to think of ways to change familiar games. In previous Acts, we have taken games played in rows and extended them to be played on rectangular boards. Lynch does the opposite. It is a variation on Checkers, played on a single long row of squares.

In his book, *The Last Recreations*, Martin Gardner introduced a game he called "Linear Checkers." It is played like checkers but all of the moves take place on a single row. Lynch is a version of linear checkers played on a row of any length. In the starting position, each player has an equal number of checkers, separated by one or more empty squares.

The game begins with pieces on opposite ends of a single row of squares. Figure 17.1 shows a simple variation in which each player has one checker on a board of length 5. Play alternates and checkers may move forward one square or jump an opponent's checker. As in checkers, you must jump if a jump is possible. The game is lost when a player loses all of her pieces. Let's call the players "black" and "white" and, as is traditional with checkers, we let black play first.

Lynch is not very interesting on short boards. White always wins 5-box Lynch with 1 checker. There isn't any strategy; every move is forced.

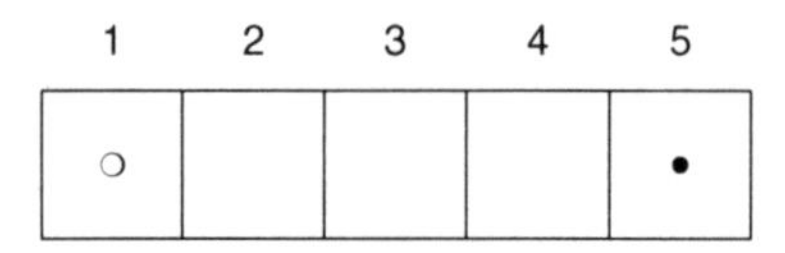

Figure 17.1

In Figure 17.1, black must play 5 to 4 and white must follow with 1 to 2. Now, in Figure 17.2, black must play 4 to 3 and white jumps and wins.

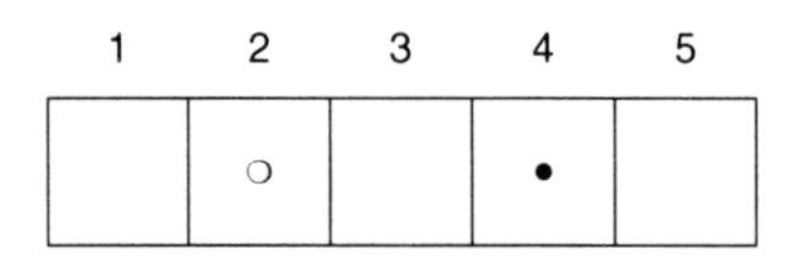

Figure 17.2

Figure 17.3 shows a similar board where each player has two checkers. Can you see that black is now forced to win?

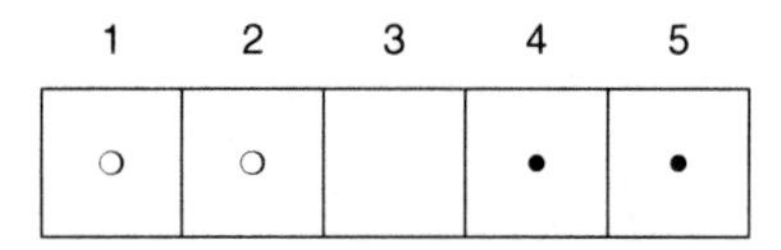

Figure 17.3

Clearly, Lynch isn't much of a game on a board of length 5.

Find a partner and play Lynch on a board of length 8, with each player having 3 checkers. Play several games until you begin to have a feel for Lynch, and then try the exercises.

Act 17 Exercises

1. Find out which player has a winning strategy in 6-box Lynch and 7-box Lynch if each player has 2 checkers.

2. When Lynch is played on longer boards, we can add more checkers. Who has a winning strategy in 3-checker, 8-box Lynch? (Figure 17.4)

1	2	3	4	5	6	7	8
•	•	•			○	○	○

Figure 17.4

3. Who has a winning strategy in 3-checker, 9-box Lynch? (Figure 17.5)

1	2	3	4	5	6	7	8	9
•	•	•				○	○	○

Figure 17.5

4. Who has a winning strategy in 3-checker, 10-box Lynch? (Figure 17.6)

1	2	3	4	5	6	7	8	9	10
•	•	•					○	○	○

Figure 17.6

Act 18

Progression: Down and Up

Progression is a placement game that, like Flit, can be played on different length boards. The name comes from the manner of play. In Going Down Progression, each move must be 1 shorter than the previous move; in Growing Up Progression, each move increases in size by 1.

Scene 1: Going Down

The game in Figure 18.1 is played on a board 15 squares in length, although it can be played on longer or shorter boards. Since $15 = 1 + 2 + 3 + 4 + 5$, the game cannot last more than 5 moves. This game began with player A covering 5 adjacent squares: [3, 4, 5, 6, 7]. Player B followed by covering [10, 11, 12, 13]. The game would continue with A playing on 3 adjacent empty squares, but in this example there is no room for A to play and so she loses.

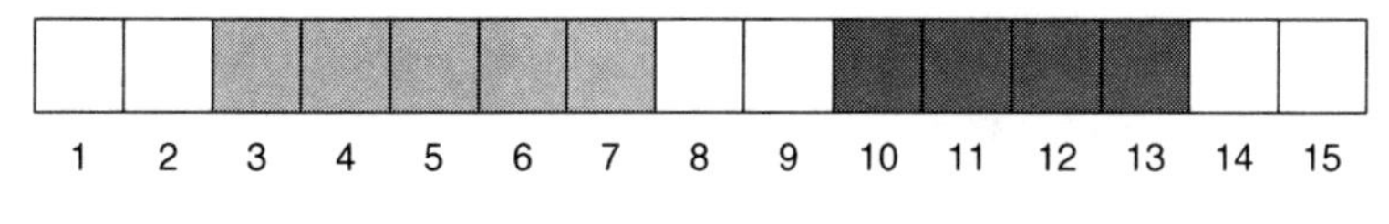

Figure 18.1

Since it is difficult to draw long boards, Progression can be played with a deck of cards. Instead of drawing 15 squares, place 15 face down cards in a row. Instead of marking adjacent squares, remove the same number of adjacent cards. Figure 18.2 shows the starting position in 10-card Progression.

Figure 18.2

As usual, the game becomes more challenging as the board increases in length. Table 18.1 lists suggested board lengths and first moves for Going Down.

Board Length	First Move
10 = 4 + 3 + 2 + 1	Cover 4 squares
15 = 5 + 4 + 3 + 2 + 1	Cover 5 squares
21= 6 + 5 + 4 + 3 + 2 + 1	Cover 6 squares
28 = 7 + 6 + 5 + 4 + 3 + 2 + 1	Cover 7 squares
36 = 8 + 7 + 6 + 5 + 4 + 3 + 2 + 1	Cover 8 squares
45 = 9 + 8 + 7 + 6 + 5 + 4 + 3 + 2 + 1	Cover 9 squares

Table 18.1

Find a partner and play Going Down several times on different length boards.

Scene 1 Exercises

1. A board of length 1 (first move: 1) is obviously a winning position, as is a board of length 3 (with first move: 2). Is a board of length 6 = 3 + 2 + 1 (first move: 3) a winning position or a losing position?
2. Who must win (with best play) on a board of length 10?

3. Figure 18.1 shows that on a board of length 15, player A loses if her first move is 3 through 7. Must player A always lose on such a board or is there a better first move that will allow her to win against any possible response by player B?

4. Suppose that a board of length $1 + 2 + 3 + ... + n$ is a losing position for player A (who must play n for her first move). Show that player A will win on the next larger board $1 + 2 + ... + n + (n + 1)$. Where should player A play her $n + 1$ block to guarantee a win?

Scene 2: Growing Up

Growing Up is similar to Going Down except that now the moves increase by 1 each time. The first player marks 1 square, the second player marks 2 adjacent squares, in the third move the player marks 3 squares, and so on. As usual, the winner is the last person to be able to move.

Since the first few moves are small, it is difficult at first to see who might have an advantage, especially on long boards.

Find a partner and play Growing Up several times on small boards (with 10 or fewer squares).

Scene 2 Exercises

1. Player A obviously wins Growing Up if the length of the board is 1. Complete Table 18.2 and decide which of the following length boards are winning positions and which (if any) are losing positions.

n	1	2	3	4	5	6	7	8
W or L?								

Table 18.2

2. Suppose A and B play Growing Up on a board of length 35. We numbered the positions 1 through 35 in Figure 18.3 for easy reference. The players filled the shaded regions—[1–8] and [11–17]—in their first 5 moves. First, player A marked 1 and B followed with [2–3]. A responded by taking [11–13] and then B played [14–17] and A played in [4–8]. What move should B play to win?

	9	10		18	19	20	21	22	23	24	25	26	27	28	29	30	31	32	33	34	35

Figure 18.3

3. In the previous problem, player B won after player A moved [4–8]. Did A have a better move available—one that would allow her to win?

4. After several moves in Growing Up, it is A's turn to play in Figure 18.4. The shaded region represents squares that have been played while the un-shaded region represents the remaining squares.

    a) If A must play a block of 10, what is the greatest possible length of the un-shaded region so that her play will be the last move?

    b) If A must play a block of n squares, how long can the un-shaded block be so that her play will be the last move?

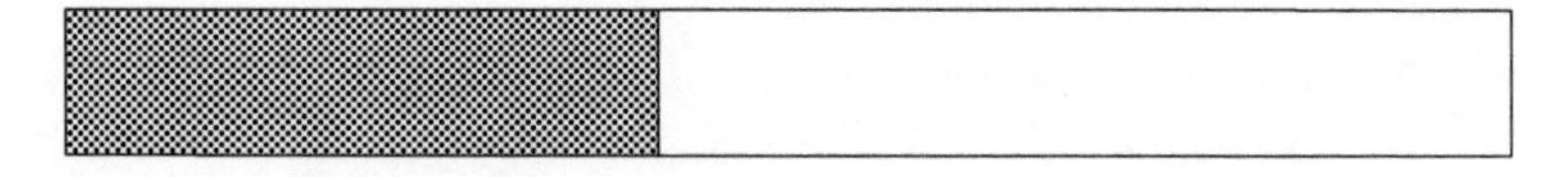

Figure 18.4

5. After several moves in Growing Up, it is A's turn to play in figure 18.4. The shaded region represents squares that have been played and the un-shaded region represents the remaining squares.

a) If A must play a block of 10, how long can the remaining block be so that she can win on the following move, no matter what B plays? What is her winning strategy?

b) If A must play a block of n squares, how long can the remaining block be so that she can win on the following move? What is her winning strategy?

Scene 3: Staircase

In this last variation, we combine the two games from Up and Down into "Staircase." We continue to play on a row of squares (or with a row of face down cards). Now, however, the move sequence is more symmetrical. In "Staircase 1, 2," player A takes 1 square, player B takes 2 squares, player A then takes 2 squares, player B takes 1 and then the pattern repeats: 1, 2, 2, 1, 1, 2, 2, 1… In "Staircase 1, 2, 3," the pattern of moves is 1, 2, 3, 3, 2, 1… and then it repeats.

Find a partner and play Staircase 1, 2 and Staircase 1, 2, 3 several times.

Scene 3 Exercises

1. Staircase 1, 2 has some features that make it like Flit or Take Away. Complete Table 18.3, showing winning and losing positions for Staircase 1, 2 on boards of length 1 through 9. The first two entries have been completed to get you started.

n	1	2	3	4	5	6	7	8	9
W/L	W	W							

Table 18.3

2. A and B's first moves are shown in Figure 18.5 in a game of Staircase 1, 2, 3. Is there a winning move for player A?

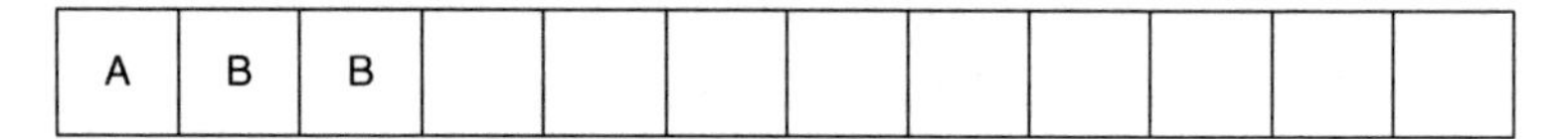

Figure 18.5

3. Now A and B's first moves are shown in this game of Staircase 1, 2, 3 in Figure 18.6. Is there a winning move for player A?

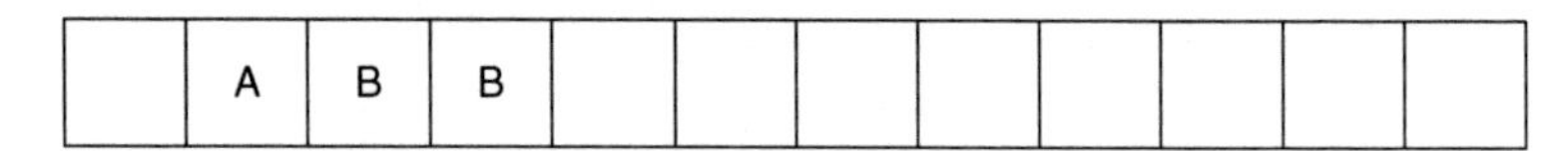

Figure 18.6

4. Figure 18.7 shows another game of Staircase 1, 2, 3 with the first moves indicated. Is there a winning move for player A?

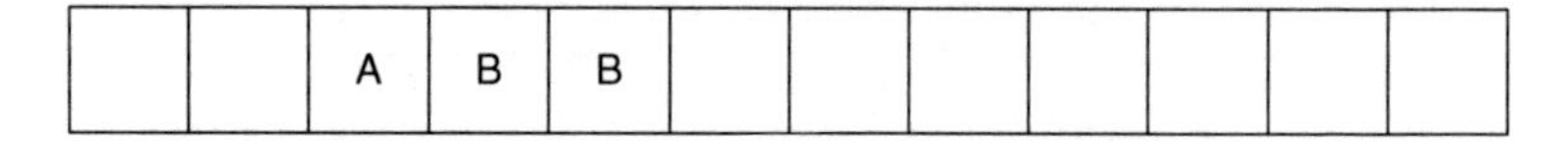

Figure 18.7

Act 19

Harder Stuff

Scene 1: Nimble

1. Which player has the advantage in the Nimble board in Figure 19.1? What strategy should that player follow to win?

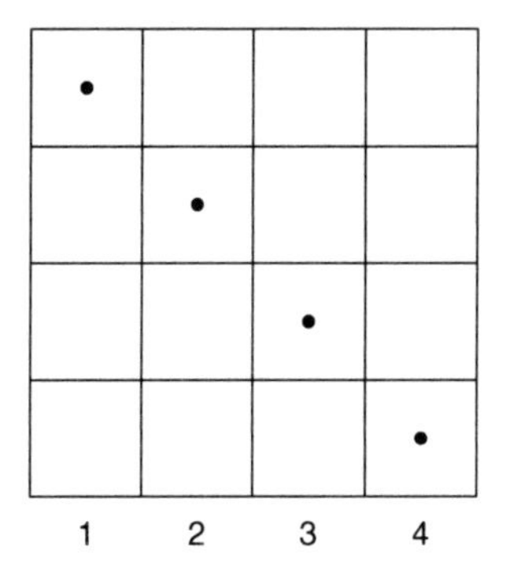

Figure 19.1

2. Which player has the advantage on the Nimble board in Figure 19.2? What strategy should that player follow to win?

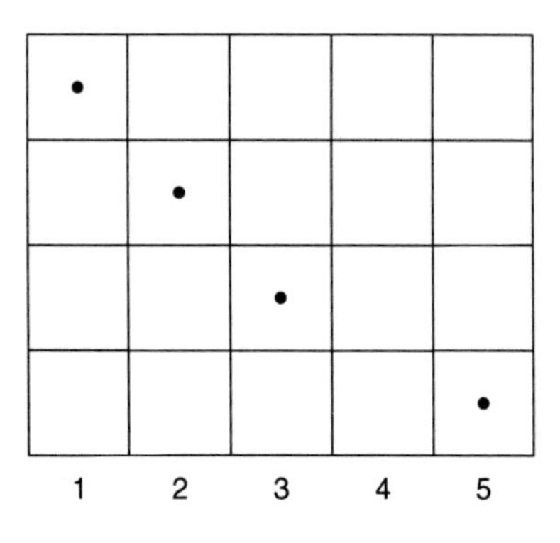

Figure 19.2

3. Which player has the advantage on the Nimble board in Figure 19.3? What strategy should that player follow to win?

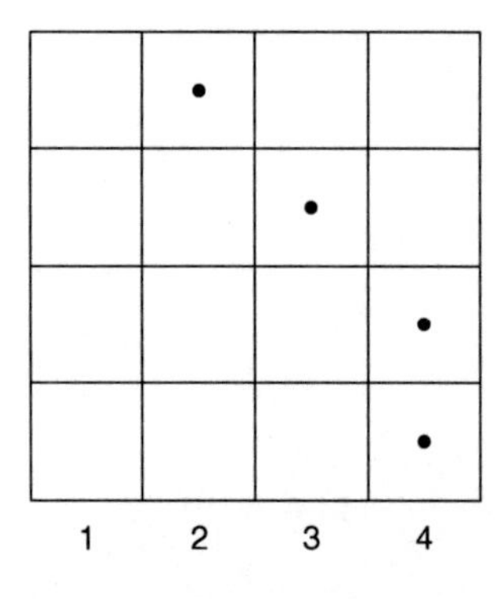

Figure 19.3

4. Which player has the advantage on the Nimble board in Figure 19.4? What strategy should that player follow to win?

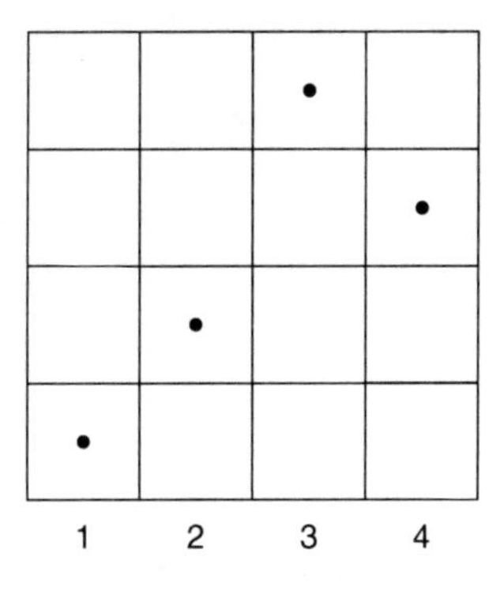

Figure 19.4

Scene 2: Variations of Take Away 1, 2, 3

1. **Variation: Choose 4**. Start with 32 counters and allow players to take 1, 2, 3, or 4 counters. Does the first person have a winning strategy? If so, what is it?

2. **Variation 2: Chaotic Take Away**. Suppose player A can take 1 or 3 counters while player B can take 1 or 2 counters. Analyze the game. Does one of the players always have a winning strategy? If not, which positions are wins for player A?

3. **Variation 3: Take and Take and Take Away**. Now both players can take 2, 3, or 7 tokens away. Can you identify the winning positions and the losing positions in this variation? What is the winning strategy?

Scene 3: Blockers

1. Blockers can be played with more than 2 tokens. Does Figure 19.5 show a winning position or a losing position in 3-token Blockers?

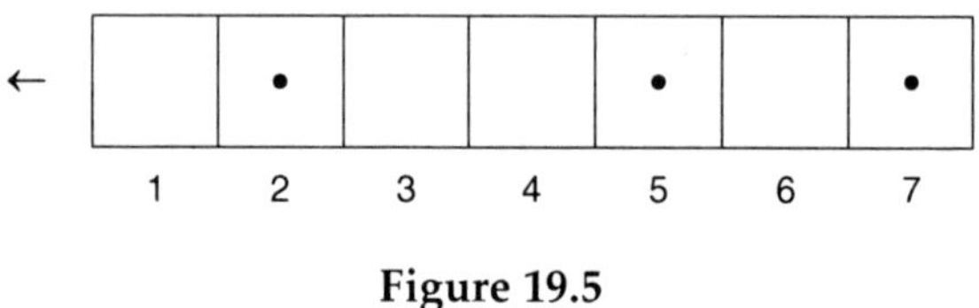

Figure 19.5

2. Figure 19.6 shows a winning position in [5, 10, 12] Blockers. Can you find the winning strategy?

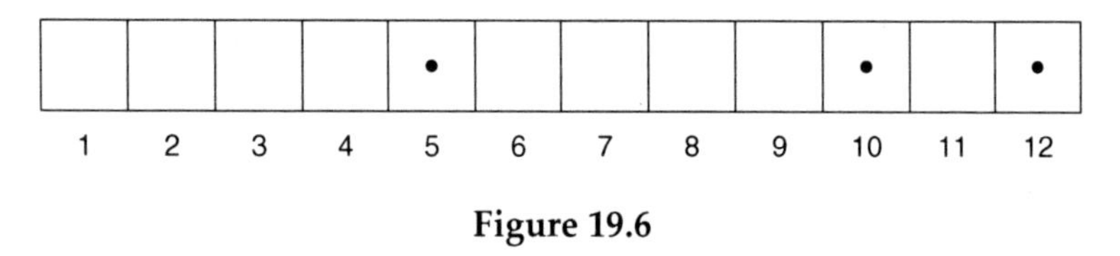

Figure 19.6

3. Find a formula for the losing positions in 3-token Blockers.

Scene 4: Take Away 1, 2, 3

1. What are the losing positions in "Take Away 1, 2, 3, 4?"
2. What are the losing positions in "Take Away 1, 2, 5?"

Scene 5: Two Piles 1, 2, 3

One way to discover the winning and losing positions of a game is to examine a game table. Table 19.1 shows possible configurations of the two piles. Several entries have already been made showing which positions are winning positions and which are losing positions.

When one pile is 0, we are really playing Take Away 1, 2, 3 and so the piles with multiples of 4 are the losing positions. But what about piles of 4 and 2, for example? Since we can take 1, 2, or 3 away from either pile, we can move from [4, 2] to [4, 1] or [4, 0] and to [3, 2], [3, 1] or [3, 0]. This corresponds to moving up 1 or 2 rows in the game table or 1, 2, or 3 columns to the left. Since we can reach [4, 0]—a losing position—so [4, 2] is marked as a winning position.

Fill out the table starting at the top left and working to the right and down.

1. When you have completed the table, see if you can find a simple way to identify all of the winning positions.

2. Find two different winning moves in [7, 12] Two Piles 1, 2, 3.

P I L E 1

P I L E 2

	0	1	2	3	4	5	6	7	8	9	10	11	12	13
0		W	W	W	L	W	W	W	L	W	W	W	L	W
1	W	L												
2	W				W									
3	W													
4	L													
5	W													
6	W													
7	W													
8	L													
9														
10														
11														
12														
13														

Table 19.1

Scene 6: Nim

1. Figure 19.7 shows a 4-token game of Nimble. Can you find the winning strategy?

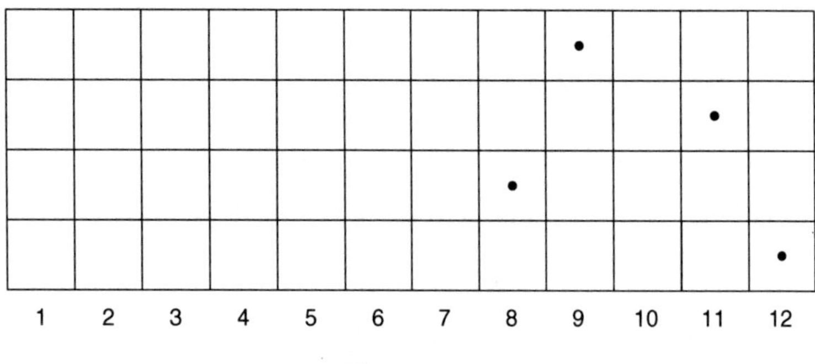

Figure 19.7

2. Figure 19.8 shows a 4-token game of Nimble. Can you find the winning strategy?

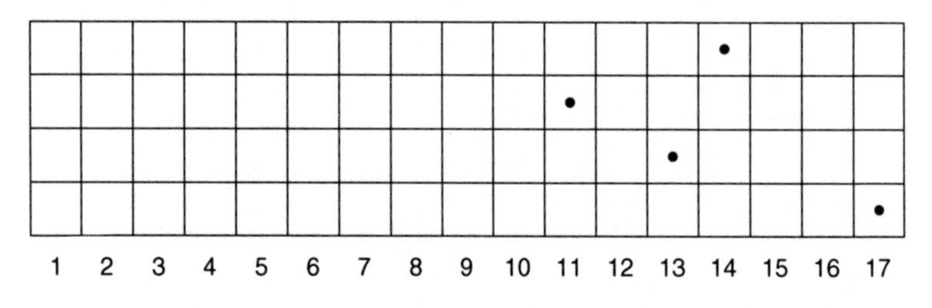

Figure 19.8

Do you see how Nim is hidden in Nimble?

3. Player A again plays Bouton's Binary Balancing Strategy but now player B decides to try something like Bouton's Strategy using powers of 3. That is, player A will divide up the piles into sub-piles of size 1, 2, 4, 8, 16, etc, whilst player B will divide up the piles into sub-piles of size 1, 3, 9, 27, etc. Remember to always start with the largest size sub-piles possible.

Fill in this table with each player's moves in [48, 32, 25] Nim. There may be times when you will have a choice of possible moves. If you have a choice, move from the smallest numbered pile.

Pile Size Before Move	Player A		Player B	
	Takes	From Pile	Takes	From Pile
48–32–25				

Table 19.2

4. You have now solved Take Away 1, 2, 3 and Two Piles 1, 2, 3 but what if we played "Three Piles 1, 2, 3?" That is, 3-pile Nim with the restriction that players can take only 1, 2, or 3 tokens from any pile. Find a winning move in Take Away 1, 2, 3 with piles of 18, 26, and 7.

5. What is a general winning strategy for Many Piles Take Away 1, 2, 3? That is, many pile Nim with the restriction that players can take no more than 3 tokens from any pile.

6. Many token Blockers contains a hidden variation of Nim. Look at Figure 19.6, which shows [5, 10, 12] Blockers. The token at 5 can move to 4, 3, 2, 1 or off the board. That is like taking 1, 2, 3 or 4 tokens from a pile. Similarly, the token at 10 can move to 9, 8, 7, or 6 since it is blocked by the first token. Moving the second token is like taking from a pile of 4 counters. In other words, if we focus on the spaces between, the tokens can then

be thought of as Nim piles. In this game, there are 3 piles of size 4, 4, and 1.

This is not ordinary Nim, however, since moving the middle token changes the number of moves for the last token. What additional rule for Nim would make the two games isomorphic?

Scene 7: Flit

1. It is hard not to notice that the game table for Flit (Figure 8.5) follows the same WWWL pattern as the game table for Take Away 1, 2, 3 (Table 4.1). This suggests that variations of Flit and Take Away may also be similar.

   You might think that enlarging the Flit board to $2 \times n$ would be similar to 2 piles Take Away, but this cannot be the case since a $2 \times n$ Flit board allows player B the copycat strategy. Instead, imagine playing two games of Flit simultaneously, each on their own rows (so that the two rows can have different lengths). Will this 2-row Flit be similar to 2 piles Take Away? To find out, complete this table for Flit played on two rows. Players try to fill each row by placing two tokens next to each other in that row.

ROW 1

ROW 2	1	2	3	4	5	6	7	8	9	10	11	12	13
1		W	W	W	L	W	W	W	L	W	W	W	L
2	W												
3	W												
4	W												
5	L												
6	W												
7	W												
8	W												
9	L												
10	W												
11	W												
12	W												
13	L												

Table 19.3

2. Can you describe the losing positions in 2-Row Flit?

Answers

Act 1 Blockers

1. A should move the token at 6 to 4.

2. This is a losing position.

3. The winning play for player A is to move to position 8. Now B can't move this token so she must move the token on 7. The winning strategy is for A to keep moving the second token until it touches the token that B moves.

4. Move the token at 40 to 18.

5. This is a winning position and A should move the token at $n + 5$ to $n + 1$.

6. The winning positions have the two tokens separated by one or more squares. The winning strategy to move the leftmost token until it is adjacent to the rightmost token. Whatever player B does now, player A continues to move against the blocking token. Eventually, player B will be forced to move the leftmost token off the board and player A will win.

Act 2 Nimble

1. Player A should move the token at 6 to square 2.

2. Move from 16 to 1.

3. Move from 16 to 9.

4. Place token on square 14.

5. The losing positions are those where two tokens occupy the same square. The winning strategy is copycat: whatever player B does, player A copies that move with the other token, keeping the two tokens together. Eventually, player A must move a token off the board and then player B will move the other token off the board to win.

6. a) 32 moves if each move is just one square.

   b) $m + n$ moves—again, each move is just one square.

7. a) Two moves. Move each token off the board.

   b) Two moves.

Act 3 More Variations

1. Move the token on 16 off the board. This leaves a losing position in 2-token Nimble.

2. [9, 9, 9] is a winning position; player A should move one token off the board, leaving [–, 9, 9]—a losing position in 2-token Nimble. The best move in [9, 9, 16] is to move the token at 16 off the board, leaving [9, 9, –].

3. Move the token at 4 off the board. This leaves [–, 10, 10], a losing two-token Nimble position.

4. This is a winning position in 2-token Nimble. Player A should move to [–, 4, 4].

5. There are many solutions. Any position with four tokens in two different columns, for example, is a winning position. If the position is [5, 12, 10, 10], for example, the winning move is to position [5, 5, 10, 10]. This new board is a combination of two losing positions in 2-token Nimble. Whether B moves

one of the tokens at 5 or at 10, player A can win by playing copycat.

6. Again, there are many possibilities. One is [5, 10, 5, 10]. Player A must move either a token at 5 or at 10 and player B can win by playing copycat.
7. $[n, n, m]$ is a winning position.
8. $[n, n, n]$ is a winning position. Player A moves one token off the board and then plays copycat.
9. $[n, n, n, n]$ is a losing position. Player B plays copycat and wins.
10. $[n, n, n, n, n]$ is a winning position. Player A moves one token off the board and then plays copycat.
11. This is a losing position. Think of it as 11 copies of $[n, n]$ Nimble.
12. $[n, n, m]$ is a winning position. Player A moves the token at m off the board and then plays copycat.

Act 4 Take Away 1, 2, 3

1. The pattern repeats as shown in Table A.1: three winning positions followed by a losing position. The losing positions are all multiples of 4.

Pile Size	1	2	3	4	5	6	7	8	9	10
Win or Lose?	W	W	W	L	W	W	W	L	W	W

Pile Size	11	12	13	14	15	16	17	18	19	20
Win or Lose?	W	L	W	W	W	L	W	W	W	L

Table A.1

2. The losing positions are piles with a multiple of 4 tokens. All other positions are winning positions.

3. Player A should remove 3 counters because 172 is a multiple of 4, a losing position.

Act 5 Two Piles: A Hidden Game

1. 11 in one pile and 15 in the other.

2. Take 5 away from the larger pile.

3. One possible rule: players take anything from either pile but the larger pile must always stay larger than the smaller pile.

4. If A takes 1, B should take 3. If A takes 2, B should take 2. If A takes 3, B should take 1. The winning strategy is to leave a multiple of 4 tokens in the pile.

Act 6 Two Piles 1, 2, 3

Scene 2

1. [18, 18] is a losing position since player B can win by using the copycat strategy.

2. Take 2 from the larger pile to make balanced piles of 85 tokens and then play copycat.

3. [24, 28] is a losing position since both piles are multiples of 4. Player B can imagine she is playing two separate games of Take Away 1, 2, 3—one with a starting pile of 24 and one with a pile of 28.

4. [88, 22] would be a winning position in Take Away since player A could just balance the two piles and play copycat. Since player A cannot take away 66 tokens, she must use a different strategy. Since 88 is a multiple of 4, A's best move is to take 2 from 22 leaving [88, 20], where both piles are now multiples of 4—a losing position. However B responds, A can play two games of Take Away 1, 2, 3 and keep both piles as multiples of 4.

Scene 3

1. Take 1 from the smaller pile leaving piles that are each multiples of 4. However B responds, A can win by playing two games of Two Piles 1, 2, 3.
2. Take 1 from the smaller pile, leaving piles that are each multiples of 4. Now follow the winning strategy in 44 Two Piles 1, 2, 3 and 64 Two Piles 1, 2, 3.
3. Player A should take 3 from the larger pile, leaving two balanced piles, and then play copycat.
4. Now player A should take 2 from the larger pile leaving $[n-2, n-2]$—equal piles.
5. A should take 1 from $4n+1$, leaving $[4n, 8n]$—piles that are each multiples of 4.

Scene 4

1. See Table A.2.

	0	1	2	3	4	5	6	7	8	9	10	11	12
0	L	W	W	W	L	W	W	W	L	W	W	W	L
1	W	L	W	W	W	L	W	W	W	L	W	W	W
2	W	W	L	W	W	W	L	W	W	W	L	W	W
3	W	W	W	L	W	W	W	L	W	W	W	L	W
4	L	W	W	W	L	W	W	W	L	W	W	W	L
5	W	L	W	W	W	L	W	W	W	L	W	W	W
6	W	W	L	W	W	W	L	W	W	W	L	W	W
7	W	W	W	L	W	W	W	L	W	W	W	L	W
8	L	W	W	W	L	W	W	W	L	W	W	W	L
9	W	L	W	W	W	L	W	W	W	L	W	W	W
10	W	W	L	W	W	W	L	W	W	W	L	W	W
11	W	W	W	L	W	W	W	L	W	W	W	L	W
12	L	W	W	W	L	W	W	W	L	W	W	W	L

Table A.2

2. In every losing position, the two piles *differ* by a multiple of 4.

Scene 5

1. a) Take 3 from 57. Think of 57 as being made of a balancing sub-pile of 22 and a different sub-pile of 35. Taking 3 away from 57 leaves two balanced piles and a multiple of 4 difference pile—a losing position for the next player.

   b) As long as player B takes from the smaller pile, player A should play copycat and take the same amount from the larger pile. Eventually, the smaller pile will vanish and player B will be forced to play in 32 Take Away 1, 2, 3—a losing position.

c) As long as player B takes from the larger pile, player A keeps the difference in the piles as a multiple of 4. In this case, A should take 1 leaving [22, 50]. Eventually, the extra multiple of 4 pile will vanish and player B will be forced to play in [22, 22] Two Piles 1, 2, 3—a losing position.

2. a) There are two winning moves. Player A can take 1 from pile 7 or 3 from pile 22. If she removes 1 from 7, there are now two balanced 6-piles and a difference pile of 16. If she removes 3 from 22, there are two balanced 7-piles and a difference pile of 12.

   b) In both cases, player B should copycat and take 2 from the larger pile—[6, 19] or [7, 16].

   c) If player B takes 3 from the larger pile, leaving [6, 19], player A can either take 1 from the larger pile or 3 from the smaller pile. If player B leaves [7, 13], then A can either take 2 from the larger pile or 2 from the smaller pile. In all cases, this leaves a difference between the piles that is a multiple of 4.

3. [25, 50] is a winning position since the difference in the piles is not a multiple of 4. Player A can win by taking 1 from the larger pile or 3 from the smaller pile.

4. Play second because [20, 32] is a losing position since the piles differ by a multiple of 4.

5. One possibility: Play two-row Nimble with the additional rule that you can only slide the tokens 1, 2, or 3 squares to the left.

Act 7 Nim

Scene 1

1. Player B wins by taking away any pile, leaving [6, 6] Nim – a balanced position. Player B will now play copycat against player A.
2. Player A should take all of the 7-pile, leaving [–, 6, 6] for player B. Since there are only two piles, Nim has become Take Away and [6, 6] is a losing position.
3. Player A should take all of the 48-pile, leaving two balanced piles for B.
4. Player A should take all of the m pile, leaving two balanced piles for B. Whatever B does, A should copycat until she takes the last counter.

Scene 2

1. Remove 5 from the 6-pile leaving [7, 1, 6] Nim.
2. If player B leaves [5, 5, 3], then player A should take all of the 3-pile. If B leaves [5, 6, 1], A should take 2 from the 6 pile.
3. [6, 7, 4] is a winning position.
4. [6, 7, 3] breaks into [(4, 2), (4, 2, 1), (2, 1)] so taking 2 from any pile is a winning move.

Scene 3

1. If we take 25 from the largest pile, we are left with [14, 5, 25, 16], which breaks into the following binary sub-piles: [(8, 4, 2), (4, 1), (16, 8, 1), (16)]. All but the 2-sub-pile are balanced. If player B uses Bouton's strategy, she should take 2 from 14, leaving [12, 5, 25, 16]—a balanced position.

2. One solution: take 23 from the 30-pile. This is like taking away a 16-pile and an 8-pile and then adding a 1-pile. This balances all of the sub-piles.

3. Take 7 from the 16-pile. This removes a 16 and adds an 8 and a 1, balancing all of the sub-piles. Another winning move is to take all of the 25-pile.

4. [1, 2, 7] is a winning position. Take 4 from the 7.

5. [25, 19, 11, 8] breaks into the binary sub-piles [(16, 8, 1), (16, 2, 1), (8, 2, 1), (8)]. There are an odd number of 8-Piles and 1-Piles so either take 9 from the 25-Pile, 9 from the 11-Pile, or 7 from the 8-Pile.

6. [37, 29, 20] breaks into [(32, 4, 1), (16, 8, 4, 1), (16, 4)]. Since the 32, 8, and 4 are unbalanced, the winning move is to take 28 (32 + 4 – 8) from 37.

7 a) Put a token on 18.

   b) Take 1 from row 1.

   c) No.

8. Move from 9 to 4 or from 10 to 7 or from 14 to 3.

9. a) Figure A.1 shows how to create a losing position by placing a token in the last row of this Nimble board.

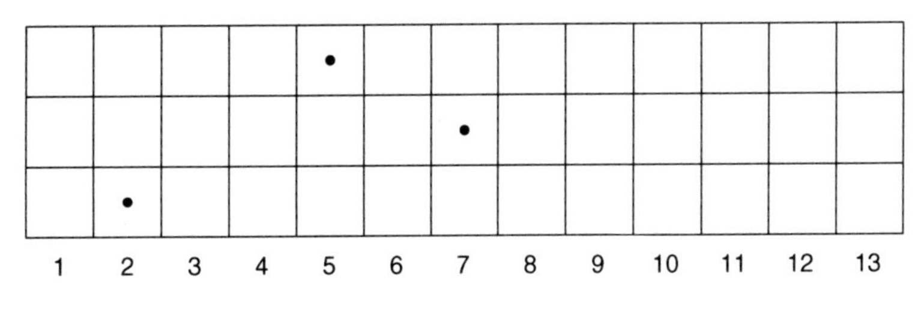

Figure A.1

   b) No. This position is the same as ([4], [4, 2], [2]). All the binary sub-piles are balanced. Any other location for the third token would leave an odd number of binary sub-piles.

Act 8 Flit

Scene 1

1. Player B must win no matter what player A does.
2. Player A can win by playing in the middle two squares.
3. Table A.3 shows the pattern WWWL repeating.

Number of boxes	2	3	4	5	6	7	8	9	10	11	12	13	14	15	16
Player A	W	W	W	L	W	W	W	L	W	W	W	L	W	W	W

Table A.3

4. Player A can win 30-box and 68-box 2Flit, but player B can win 81-box 2Flit since that is a losing position.
5. Even boards are wins for player A. Strategy: take the center squares and play copycat. Because an even board is symmetric about its center, if B can move there will be a mirror image location for A to move.
6. $n = 4k + 1$ where k is any whole number.

Scene 2

1. Player A should play in boxes 49, and 50 leaving 48 boxes on either side of her move. Wherever B plays there will be a symmetric move available for A so that A can win by playing copycat.
2. A should play in boxes 1 and 2 or in boxes 42 and 43. In each case, B has to play in a region of length 41—a losing position.
3. A's move divided the board into two regions with 24 boxes to the left and 15 boxes to the right of her move (Figure A.2). Since 24-box Flit is a losing position, B should seize the center and play in boxes 12 and 13.

4. a) If A plays in boxes 47 and 48, where should B play? Player A's move created a region of six boxes and a region of five boxes. A should play in the center of the even region in boxes 43 and 44. (Figure A.2)

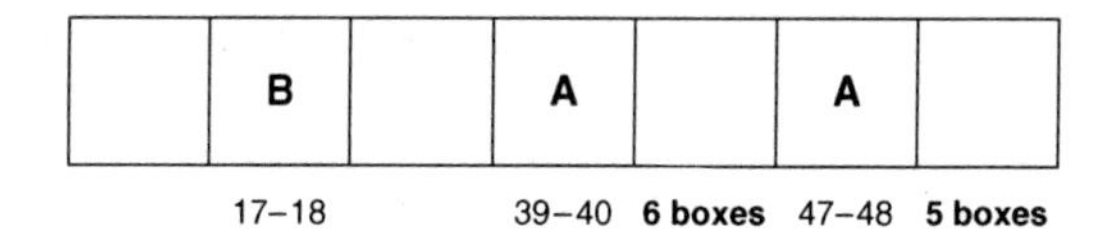

Figure A.2

b) Now A has created a region on the left of length 7 and a region on the right of length 6. Once again, B should play in the center of the even region in boxes 12 and 13.

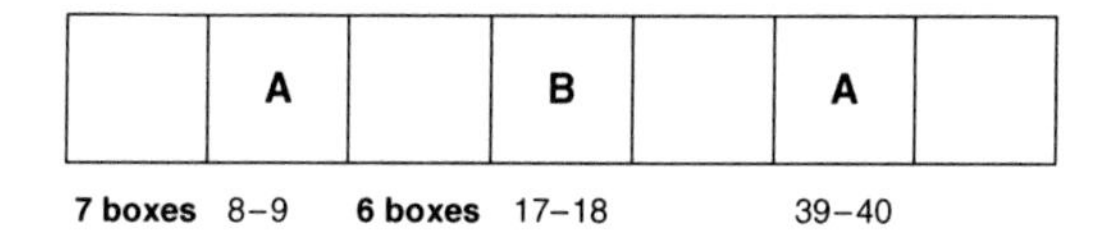

Figure A.3

Act 9 Mr Flit

Scene 1

1. The board is symmetrical so player B can win by playing copycat.

2. Player B can always win on a 3×3 board. Player A's move will either include a corner square or not. In either case, B can play next to A to complete a 2×2 filled square, leaving two moves around it.

Scene 2

1. The winning move for player A is [b3, c3]—the two center squares (Figure A.4). Now whatever player B does, A can balance by playing copycat.

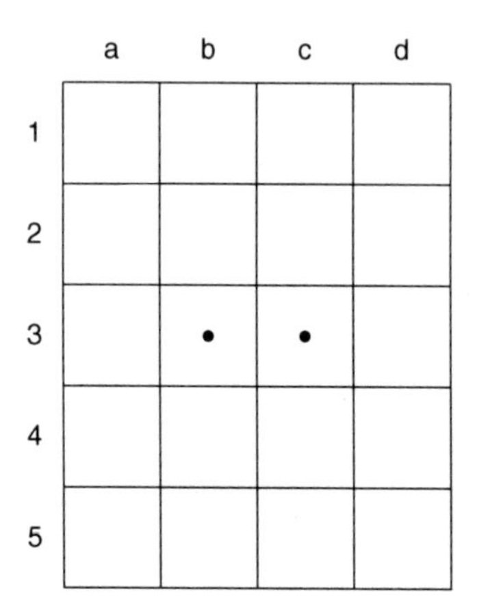

Figure A.4

2. Row 7 divides the board in half and player A should play in the center of this row, in squares 12 and 13—the two center squares on the board. Whatever player B does, player A now has a symmetrical move in response. A wins by playing copycat.

3. There is no central row or column on this board. There is, however, a central point of symmetry where squares 10 and 11 in rows 6 and 7 touch. Wherever A plays, there will be a symmetrical place through this center for B to play. B wins by playing copycat.

4. Since there are an odd number of columns, there is a central column with an even number of squares. Player A should move into the center two squares of this column and then play copycat. Player A has the advantage by using symmetry.

5. There is no central move for player A to claim so wherever A moves, B can win by playing in the symmetrical position through the center.

6. The problem with an odd×odd board is that there is neither a center point nor a center row or column. Instead, there is a central square. Since players move in twos, player A cannot take the center of the board and player B cannot always copycat A. If player A avoids that central square, then player B will have a copycat place to play. But if A plays in the center, there is no symmetrical place for B to play and so neither player can take advantage of symmetry. Figure A.5 illustrates this situation with a 5 × 7 board. Player A began in [c3, c4]. Since there was no symmetrical move available, B played [c5, d5]. The board now looks fairly random. There may be a win here for player A, but it is not by using a simple balancing strategy.

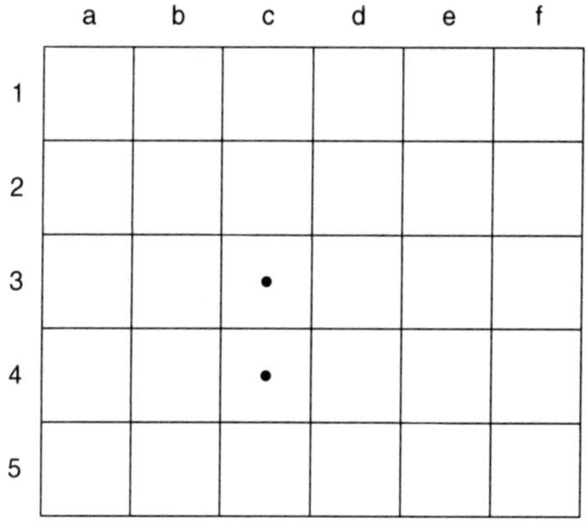

Figure A.5

7. a) An odd × odd board has a center square. Player A should play one token in the center. She can now win by playing copycat. Wherever player B moves and whether B plays one or two tokens, A will have a symmetrical move.

   b) An odd × even board has either a center row or center column with an even number of squares. Player A should play 2 tokens on the center two squares. A will win if she plays a copycat strategy.

c) An even $\times$ even board has no central square, but there is a central point of symmetry. Wherever A plays, B will win by playing copycat.

8. Since a move consists of placing three tokens, all moves are horizontal and so whether A plays in row 1 or row 2, player B can play in the same squares in the other row, balancing A. Therefore, B will always win on a $2 \times n$ board.

9. Player A can always win on a $3 \times n$ board if n is odd. There is a central column and A should move vertically in that column. This will divide the board into two symmetrical halves, and now A can win by playing copycat. If n is even, there is no center column and the board is symmetrical around a center point. If A avoids that center then B will always be able to copycat A. Therefore, A's best move is to play in the middle row, covering the point of symmetry. This breaks the overall symmetry of the board and gives A the best chance of winning.

10. a-d) All rings are symmetrical, so player B can always win by following a symmetry strategy.

    e) It doesn't matter whether players can place 1 or 2 tokens since B can still win by playing copycat.

Act 10 Landis

1. A wins by playing in column b.

2. Player A can win by playing [d3, c3, c4]. This leaves no room for B to play.

3. Eight moves. One solution is shown in Figure A.6.

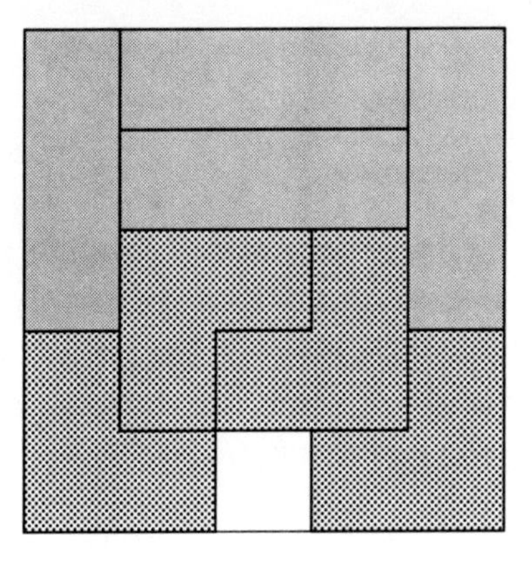

Figure A.6

4. The shortest game we have found is 5 moves. Our solution is shown in Figure A.7.

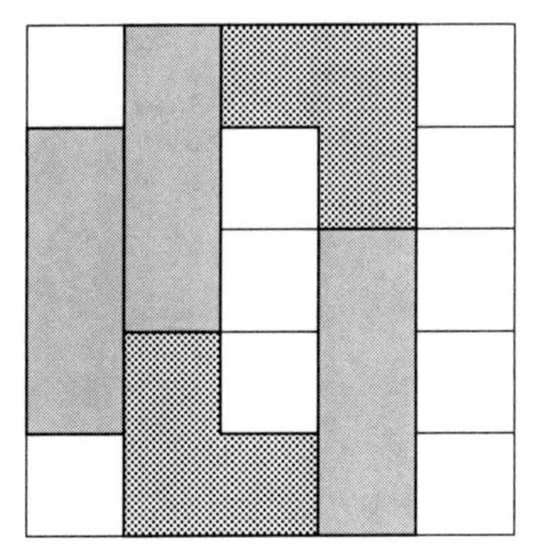

Figure A.7

5. The I-player should cover square a4 or a5 or both. Any other move will give L an extra move and the game.

6. a) There are 16 different first moves. However, by turning the board, all eight corner plays are equivalent, as are each of the eight non-corner plays. In that sense, there are only two possible first moves.

b) Since there is a central point of symmetry, player B can win by playing copycat.

7. a) 36 different first moves.

   b) Since the board is symmetrical through its center point, player B can copycat any move player A makes, *except one that wraps around the center*. If A plays [b2, b3, c3] for example, B cannot play copycat. This makes it more difficult for B to find a win and so it is A's best move.

Act 11 Add 'em Up

1. The five possible games are:

   8 8 8 → 1 6 8 → 7 8 → 1 5 → 6

   8 8 8 → 1 6 8 → 1 1 4 → → 2 4 → 6

   8 8 8 → 1 6 8 → 1 1 4 → 1 5 → 6

   8 8 8 → 8 1 6 → 9 6 → 1 5 → 6

   8 8 8 → 8 1 6 → 8 7 → 1 5 → 6

2. Player B must win because there is always an even number of moves.

3. 4 6 8 → 1 0 8 → 1 8 → 9

   4 6 8 → 4 1 4 → 5 4 → 9

   4 6 8 → 4 1 4 → 4 5 → 9

4. a) 1 2 3

   b) The three digits' total is less than 10.

5. 1 1 1 1 will end in 3 moves.

   1 1 1 1 1 will end in 4 moves.

1 1 1 1 1 1 will end in 5 moves.

n 1s will end in $n-1$ moves only if $n<10$. If $n \geq 10$, then there will be extra moves possible because the 1s will sum to a 2-digit number.

6. a) 8 8 8 → 1 6 8 → 7 8 → 1 5 → 6 Player A wins. The score is 31 to 13.

   8 8 8 → 1 6 8 → 1 1 4 → 2 4 → 6 Player B wins. The score is 18 to 20.

   8 8 8 → 1 6 8 → 1 1 4 → 1 5 → 6 Player A wins. The score is 21 to 20.

   8 8 8 → 8 1 6 → 9 6 → 1 5 → 6 Player A wins. The score is 31 to 15.

   8 8 8 → 8 1 6 → 8 7 → 1 5 → 6 Player A wins. The score is 31 to 13.

   b) If Player A makes 8 6 7 → 8 1 3, she may be able to score as much as 25. If A plays 8 6 8 → 1 4 7, the best she can do is score 16 against a good opponent.

7. The best scores we have found so far are:

   154: 1 2 3 4 5 6 7 8 9 → 1 2 3 4 5 6 7 1 7 → 1 2 3 4 5 6 7 8 → 1 2 3 4 5 6 1 5 → 1 2 3 4 5 6 6 → 1 2 3 4 5 1 2 → 1 2 3 4 5 3 → 1 2 3 4 8 → 1 2 3 1 2 → 1 2 3 3 → 1 2 6 → 1 8 → 9

   216: 9 9 9 9 9 9 9 9 9 → 1 8 1 8 1 8 1 8 9 → 9 9 9 9 9 → 1 8 1 8 9 → 9 9 9 → 1 8 9 → 9 9 → 1 8 → 9

   143: 9 8 7 6 5 6 7 8 9 → 1 7 7 6 5 6 7 1 7 → 8 7 6 5 6 7 8 → 1 5 6 5 6 1 5 → 6 6 5 6 6 → 1 2 5 1 2 → 3 5 3 → 8 3 → 1 1 → 2

8. The highest total requires using the longest string of digits. Therefore, we choose either to add $8+87766557788$ or $88776655778+8$. The latter sum is greater.

Act 12 Connect-the-Dots

1. There are 9 rows with 8 lines in each row for a total of 72 horizontal lines. There are another 72 lines in the 9 columns—144 lines altogether.

2. a) There are only 4 lines. Since play alternates, the second player makes the last move and completes the only square.

   b) No. The first player can win against a careless opponent, but most games will produce a tie.

3. Play a line in the lower left-hand corner. Player B will have to give up a square. If she plays on the right side of the board, you win a square; and then, since you play again, you must give up a square too. Your second move should give up the other square on the right side of the board. Player B will get a point but will then have to draw a line on the left side of the board and you will win both squares there.

4. Player B should win.

5. This game has not been fully analyzed, but we think the first player should win.

Act 13 Boxes

1. If player A surrounds the board with her box, there will be no room for player B to make any move at all.

2. The same strategy works on a $2 \times n$ board; player A should play around the entire board.

3. On a $3 \times n$ board, A should not play around the entire board, since that will leave one or more squares in the center where B can play. Instead, player A should surround the middle row or surround two rows as illustrated below in a 3×11 game in Figure A.8. Player B has no move at all.

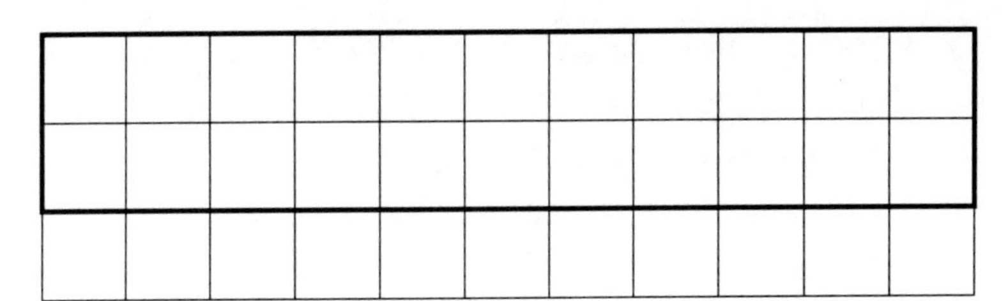

Figure A.8

4. Player B can win with a 2×2 box in the center of the board.

5. a) Player B can win with a 2×2 box in the lower-right corner. Since the board is now symmetrical, any move that player A makes will leave a symmetrical move for player B.

   b) If player B plays either a 1×1 box in the lower-right corner or a 1×1 box, touching the lower-right corner of the first move (Figure A.9), player A can respond by taking the other 1×1 box located along the diagonal, leaving a position shown in Figure A.9. Now player A can win by following the same symmetry strategy outlined in (a).

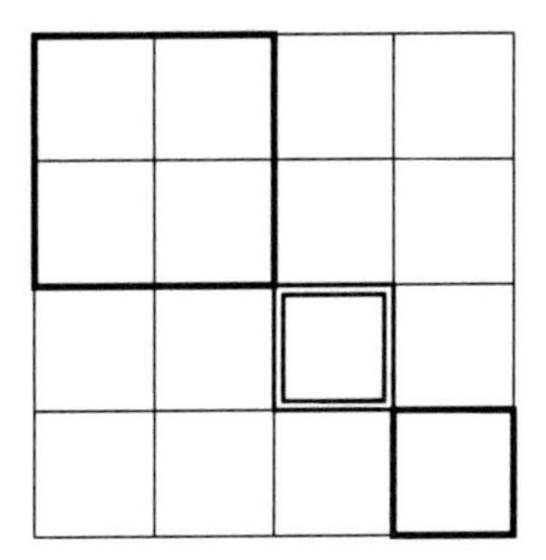

Figure A.9

6. a) Player A should surround either a4 or b4. If A takes a4 then player B can only take c4 or d1, leaving the last move for A. If A takes b4, then B can take either a3 or d1 and again, A wins.

   b) If player A takes [a4, b4], player B wins with d1.

7. If player A plays around the 2×2 box [f2, g2, f3, g3], then player B can win by playing around [j1, j2, j3, j4].

8. Player B can win by leaving two moves—one for A, and the last for her. Since g3 is the only move available inside the large box, B must make sure that there are two moves in the right-most column. B can win by playing j1, j2, j3 or j4. Any of these moves leaves a second move in column j as well as g3.

9. a) 50

   b) What are the fewest number of moves possible on a 10×10 board?

10. a) The greatest number of moves on an odd-sided square $(2n-1) \times (2n-1)$ is $2n^2 - 2n + 1$. The greatest number of moves on an even-sided square $2n \times 2n$ is $2n^2$.

    b) The fewest number of moves possible on a $(2n-1) \times (2n-1)$ square is n. The fewest number of moves possible on a $2n \times 2n$ is also n.

Act 14 Hold That Line

1. Player A is forced to win in three moves. Whichever dots she connects, B must leave a move to end the game.

2. There are 36 possible first moves in 3×3 Hold That Line.

3. Player A should join c3 to a2 so that player B can win by connecting a2 to b2.

4. Join c1 to b1.

5. 8 moves are possible with cooperation. The shortest game has only 3 moves. Can you find them?

6. Player A should begin with either a horizontal, vertical, or diagonal line through the center dot. This divides the board in half. Player A can now win by following a symmetry strategy.

Whatever player B does there, will be a symmetrical move on the other side of the board available to player A. Eventually, player A will take the last move.

7. The same strategy for 3×3 Hold That Line will work in 4×4 Hold That Line as well, but now the first move must be along a diagonal since there isn't a central row or column of dots.
8. The symmetry strategy will work on any square board. In fact, it will work on any rectangular board. Player A need only begin by joining two opposite corners along a diagonal.
9. Symmetry will still work in this variation. The first player can always win.
10. The same symmetry strategy will continue to produce a win for the first player.

Act 15 The Fifteen Game

1. Player A can win by playing 5. Player B must respond with 1 to block. Player A should respond with 2 or 4, threatening to win with either [2, 5, 8] or [2, 9, 4]. Since B cannot block both threats, A will win.
2. Player A is correct. She is threatening to win so B must play 5 to block. Since B now threatens to win with [8, 5, 2], A must play 2 to block B's threat. But now, A is threatening to win in two different ways. 6 + 2 needs 7 to win and 2 + 4 needs 9 to win. B cannot block both threats and will lose.
3. A can play 6, forcing B to play 5. A then plays 8 and again has two threats and so will win. Alternatively, A can play 3, forcing B to play 8 to block. Now A plays 5 and again has two different winning threats.
4. Player A can play 2, 4, 6 or 8. Each of these forces B's response. Whatever B does, A can create a double threat. For example,

if A plays 8, then B must block with 2. A's move of 6 is forced to stop B from winning and now A will win with either 1 or 4. B cannot block both threats.

5. There are many ways to tie this game. One possible sequence is: [A, 5], [B, 2], [A, 7], [B, 3], [A, 6], [B, 4], [A, 9], [B, 1], [A, 8].

6. a) The numbers in each row, each column, and the two main diagonals all sum to 15.

   b) Tic-Tac-Toe.

   c) X and O.

7. If we think of Tic-Tac-Toe, it is much easier to see that A can win by playing in any corner square (that is, choosing 2, 4, 6, or 8). If A plays 8, for example, B must block with 2, preventing a win along the diagonal. But now A can play in the middles of the left edge (1), threatening to win along both a row and column ([6, 1, 8] or [1, 5, 9]).

8. Player A has several options. One is to play 2, forcing 4 in response, then 5 creates a double threat.

Act 16 Sliders

1. Player B must win. Clearly, player A should not remove an entire row or column on the first move or player B could win immediately. So player A must remove just one token from some corner, leaving a position similar to the left side of Figure A.10. In this case, player B wins by removing the opposite corner piece leaving a winning position like the one on the right.

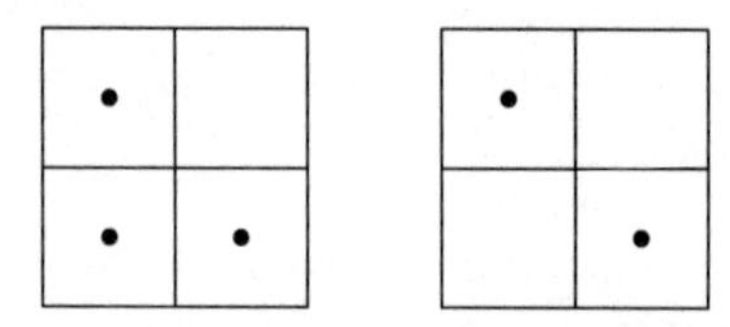

Figure A.10

2. Player A should slide the bottom row to the left, leaving the winning position shown in Figure A.11.

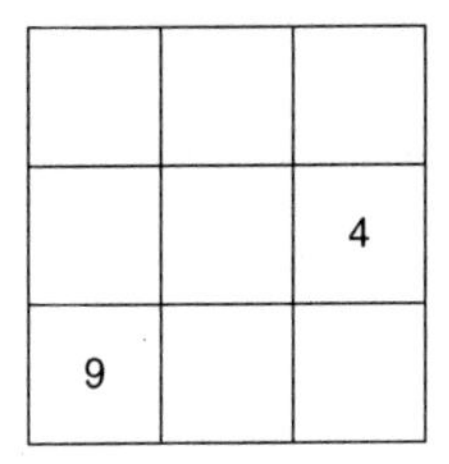

Figure A.11

3. Yes. Slide the bottom row to the right one square, leaving the 2×2 square shown in Figure A.12. As in problem 1, this is a losing position.

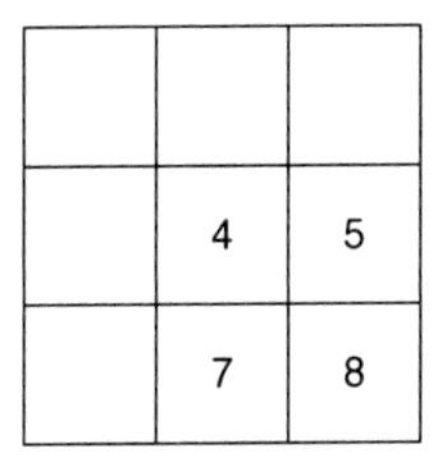

Figure A.12

4. This is a winning position if the next player removes the middle row. There are 2 moves left.

5. The symmetry strategy will not work here because some moves will destroy the possibility of symmetry. For example, the move shown in Figure A.13 has no symmetrical response.

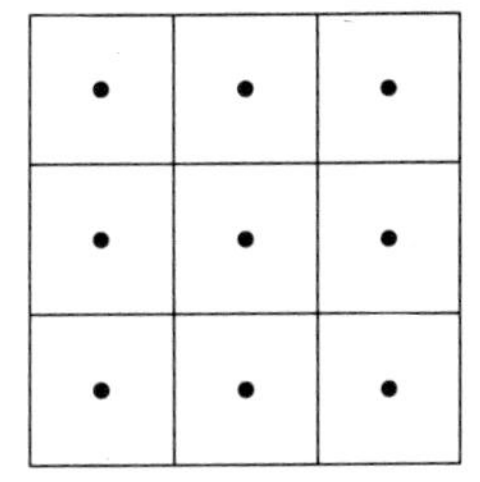
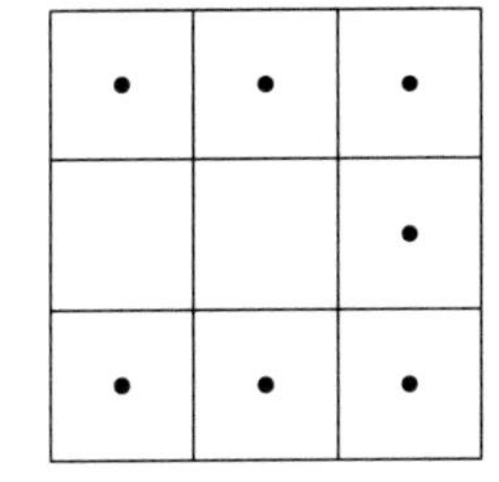

Figure A.13

Act 17 Lynch

1. Black wins in 6-box, 2-checker Lynch and white wins 7-box, 2-checker Lynch.

2. White wins 3-checker, 8-box Lynch.

3. Black wins 3-checker, 9-box Lynch.

4. Black can win. All the moves are forced (assuming neither player gives up checkers voluntarily) until the position reached in A.14. If black moves 3 to 4, white will win but if black moves 1 to 2, black will win.

1	2	3	4	5	6	7	8	9	10
•		•		•		○	○	○	

Figure A.14

Act 18 Progression

Scene 1

1. When $n = 1$ (first move: 1), the win is automatic. When $n = 3$ (first move: 2), player A must lose; no matter where she plays, there will be a move for B. When $n = 6$ (first move: 3), Player A can assure a win by playing in the first three or the last three squares. In either case, the board now has 3 squares and however player B takes 2, there will be a last square for A to take.

2. Player B has the win. There are only 4 different places A can move, as shown in Figure A.15. If B plays 3 squares adjacent to

A's, there will always be space for A to take 2 and B to claim the last square and win.

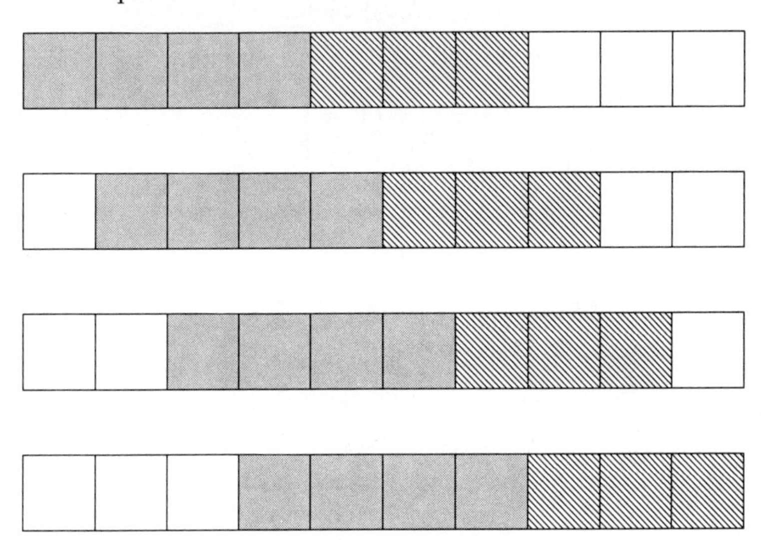

Figure A.15

3. If A plays in [4, 5, 6, 7] (Figure A.16), then B must play somewhere in [9, 15]. A can now play in [1, 2, 3] and B cannot prevent 2 more moves in the game.

1	2	3	4	5	6	7	8	9	10	11	12	13	14	15

Figure A.16

4. Player A should take the first $n + 1$ squares (or the last $n + 1$ squares). In each case, this leaves B to play onto a board that we are told is a losing position.

Scene 2

1. Table A.4 shows the completed values.

n	1	2	3	4	5	6	7	8
W or L?	W	W	W	L	L	L	W	W

Table A.4

2. B should take squares 24 through 29, leaving 2 blocks of 6 squares so that A cannot play.

3. Player A should have taken any block of length 5 within [18, 35]. No matter where B takes her block of 6, there will be a block of length 7 for A and there will not be a block of 8 for B.

4. a) You might think that 20 is the longest, since A would leave a block of 10 for B, but the correct answer is 30. If A takes the middle 10 squares there will be no room for B to play.

   b) $3n$ following the logic described in (a).

5. a) When player A moves, she will take 10 squares, dividing the un-played region into two parts. She wants to insure that she can play next, so she must reserve 12 squares for her future play. If she makes one part have 12 squares and the other 36 squares, then no matter how B takes 11, A will be able to take 12 and win. Therefore, the longest possible block of un-played squares is $12 + 10 + 36 = 58$ squares and A should remove a block that will leave one block of 12 and another block of $3(12) = 36$.

   b) Player A will take n squares, dividing the un-played region into two parts. If one part has $n + 2$ squares and the other has $3(n + 2)$ squares, then no matter how B takes $n + 1$ squares, A will be able to take $n + 2$ squares and win. Therefore, the longest possible block of un-played squares is $(n + 2) + n + 3(n + 2) = 5n + 8$ squares and A should remove a block that will leave one block of $n + 2$ and another block of $3(n + 2) = 3n + 6$ squares.

Scene 3

1. When $n = 2$ or 3, the winning move is to play 2. When $n = 5$, player A should play 3. When $n = 7$, player A should take square 1, since that leaves player B with a losing position. When $n = 8$, the winning move is in square 4 and when $n = 9$ it is square 2. These results are summarized in Table A.5.

n	1	2	3	4	5	6	7	8	9
W/L?	W	W	W	W	W	L	W	W	W

Table A.5

2. No. This is a losing position and any move that A makes will allow B to win.

3. No. This is a losing position and any move that A makes will allow B to win.

4. Yes. The winning play is shown in Figure A.17.

		A	B	B			A	A	A		

Figure A.17

Act 19 Harder Stuff

Scene 1: Nimble

1. Player A can win by moving the token on square 4 off the board. Can you figure out how?

2. Player A can win by moving the token at 4 off the board.

3. Player A can win by moving the token at 3 to 2 and then thinking of the board as [2, 2] Nimble and [4, 4] Nimble. The winning strategy is to copycat player B's move.

4. Although the board looks different, this is the same game as in (2), so moving the token at 4 off the board is the winning move.

Scene 2: Variations of Take Away 1, 2, 3

1. **Choose 4**. Now the winning strategy is to keep the pile a multiple of 5. Therefore, player A should take 2 counters, leaving a multiple of 5 tokens. A should always maintain a multiple of 5 tokens in the pile so that eventually player B will have to play into a pile of 5 and lose.

2. **Chaotic Take Away**. Table A.6 is a game table of Chaotic Take Away, showing whether a position is a winning or a losing one for player A.

	0	1	2	3	4	5
0	L	W	L	W	L	W
1	W	L	W	L	W	L
2	L	W	L	W	L	W
3	W	L	W	L	W	L
4	L	W	L	W	L	W
5	W	L	W	L	W	L

Table A.6

It should be clear that player A loses if both piles have an even number of tokens or both have an odd number of tokens.

3. **Take and Take Away**. Again, a game table will help to show the possibilities (Table A.7). Remember, the first player who cannot move is the loser. Since neither player can remove one token, a board with one token is a losing position.

=	0	1	2	3	4	5	6	7	8	9
0	L	L	W	W	W	L	L	W	W	W
1	L	L	W	W	W	L	L	W	W	W
2	W	W	L	L	W	W	W	L	L	W
3	W	W	L	L	W	W	W	L	L	W
4	W	W	W	W	L	W	W	W	W	L
5	L	L	W	W	W	L	L	W	W	W
6	L	L	W	W	W	L	L	W	W	W
7	W	W	L	L	W	W	W	L	L	W
8	W	W	L	L	W	W	W	L	L	W
9	W	W	W	W	L	W	W	W	W	L

Table A.7

It is quite difficult to describe the losing positions even though you may see the pattern in the game table. The losing positions lie along diagonal lines—two 2×2 square blocks followed by a single block of 1. This pattern then repeats.

One way to describe the pattern is to say that when the piles are divided by 5, there are three types of losing positions. Either the two piles have the same remainders, or if one pile has remainder 0, then the other has remainder 1, or if one pile has remainder 2, then the other pile has remainder 3.

The winning strategy, as always, is to move so that your opponent has a losing position.

Scene 3: Blockers

1. Figure A.18 shows a winning position. The winning move is from 2 to 1.

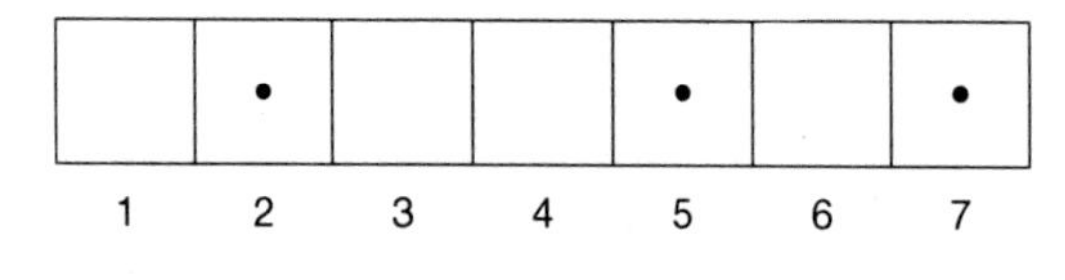

Figure A.18

If player B moves token 1 off the board, then player A can use the winning strategy for 2-token Blockers. If player B moves the token on 5, then play copycat with the token on 7, always leaving 1 square between those tokens. Eventually, player B will have to move the token off the board and now player A can play 2-token Blockers.

Finally, if player B moves the token on 7, play copycat with the token on 5. Eventually, the three tokens will occupy squares [1, 2, 3]—a winning position for player A.

2. Figure A.19 shows the board. The winning move is from 5 to 1, leaving the position [1, 10, 12]. If player B moves 1 off the board, then A wins by playing 2-token Blockers. If B moves 12 to 11, then A wins by removing the token at 1, leaving a losing position in 2-token Blockers. Finally, if B moves the middle token, then A copycats with the rightmost token, always

keeping one space between them. Eventually, this will force B to play in [1, 3, 5] Blockers where all moves lose.

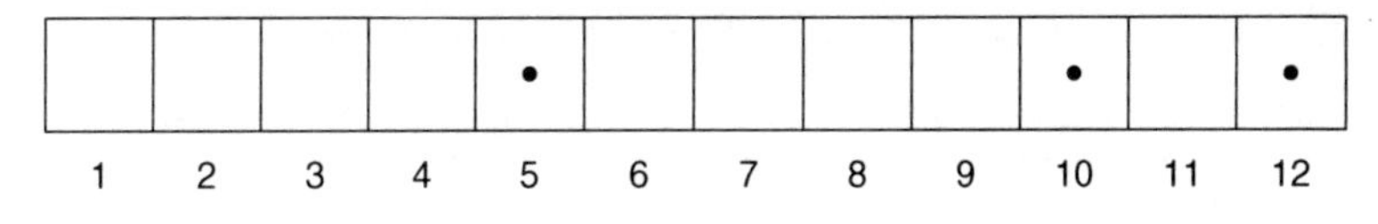

Figure A.19

3. Call the positions of the tokens l, m, and r for left, middle and right. The losing positions are the ones where $r = m + l + 1$—that is, where the two rightmost tokens are separated by one more square than the position of the leftmost token.

Scene 4: Take Away

1. Multiples of 5 are the losing positions.

2. Table A.8 shows the pattern of losing positions in Take Away 1, 2, 5.

1	2	3	4	5	6	7	8	9	10	11	12	13	14	15	16
W	W	L	L	W	W	L	W	W	L	W	W	L	W	W	L

Table A.8

The losing positions are 3 and—starting with 4—any position that is one more than a multiple of 3.

Scene 5: Two Piles Take Away 1, 2, 3

We completed the Game Table A.9 by noticing that taking 1, 2 or 3 away from a pile means moving a column up or to the left in a row by 1, 2 or 3 squares. For example, [6, 10] Take Away 1, 2, 3 is a losing position because moving 1, 2, or 3 either up or to the left gives a winning position.

PILE 1

PILE 2

	0	1	2	3	4	5	6	7	8	9	10	11	12	13
0	L	W	W	W	L	W	W	W	L	W	W	W	L	W
1	W	L	W	W	W	L	W	W	W	L	W	W	W	L
2	W	W	L	W	W	W	L	W	W	W	L	W	W	W
3	W	W	W	L	W	W	W	L	W	W	W	L	W	W
4	L	W	W	W	L	W	W	W	L	W	W	W	L	W
5	W	L	W	W	W	L	W	W	W	L	W	W	W	L
6	W	W	L	W	W	W	L	W	W	W	L	W	W	W
7	W	W	W	L	W	W	W	L	W	W	W	L	W	W
8	L	W	W	W	L	W	W	W	L	W	W	W	L	W
9	W	L	W	W	W	L	W	W	W	L	W	W	W	L
10	W	W	L	W	W	W	L	W	W	W	L	W	W	W
11	W	W	W	L	W	W	W	L	W	W	W	L	W	W
12	L	W	W	W	L	W	W	W	L	W	W	W	L	W
13	W	L	W	W	W	L	W	W	W	L	W	W	W	L

Table A.9

1. The losing positions are those where the difference of the piles is a multiple of 4.
2. Take 1 from the 12-Pile or 3 from the 7-Pile.

Scene 6: Nim

1. [8, 9, 11, 12] Nimble is isomorphic to [8, 9, 11, 12] Nim. Use Bouton's strategy and remove 2 from the 12-pile by moving the token in row 4 to square 10.
2. Bouton's strategy tells us to remove 17 from pile 4 and add back 8. The winning move, then, is to move the token in the

last row to square 8. Nim is hidden in Nimble, and vice-versa, because the number of squares a token can move in Nimble is like the number of tokens in a Nim pile.

3. The moves in [8, 12, 25] when player A balances powers of 2, whilst player B balances powers of 3, as shown in Table A.10. B loses.

4. The winning move is to take 1 away from pile 18, 1 away from pile 26 or 3 away from pile 7.

Pile Size Before Move	Player A		Player B	
	Takes	From Pile	Takes	From Pile
48–32–25	9	25		
48–32–16			10	16
48–32–6	10	48		
38–32–6			6	38
32–32–6	6	6		

Table A.10

5. Consider the previous problem where we were playing 3-Pile Take Away 1, 2, 3 with piles of 18, 26 and 7. If we were playing Nim, we would break these piles into binary sub-piles of [(16, 2), (16, 8, 2), (4, 2, 1)]. Since the 8, 4, 2, and 1 are all unbalanced, we would take away $8+2-4-1=5$ from pile 2 if we were playing Nim. But we cannot take away more than 3 from any pile. Think again about the unbalanced piles of 8 and 4. Are they really problems? Since we are taking away 1, 2, or 3, multiples of 4 are losing piles and so sub-piles of 8 or 4 are already

losing piles. They do not need to be balanced. Only the 2 and 1 piles are unbalanced. They can be balanced by taking $2-1=1$ from the 18 or the 26 or $2+1=3$ from the 7.

In general, we break the piles into binary sub-piles but we can ignore all the sub-piles except the 2s and 1s, since they are multiples of 4. If there is an unbalanced 2 or 1, the winning move is to balance it.

To find the winning strategy, imagine a losing (balanced) position. A player taking 1, 2 or 3 will cause the 1-piles or the 2-piles (or both) to become unbalanced. The winning move is to rebalance them.

6. Here is one possible rule: When counters are removed from pile 1, they are added to pile 2. When tokens are removed from pile 2, they are added to pile 3, and so on. The only way to reduce the total number of counters is to move the rightmost token. With this additional rule, this Nim game is the same as Blockers.

Scene 7: Flit

Table A.11 shows the completed game table. As in Two Piles 1, 2, 3, positions that differ by a multiple of 4 are losing positions. However, there are other losing positions as well, forming the opposite corners of each 2×2 square. These can best be described algebraically. If m and n are any whole numbers, then the other losing positions have $2+4m$ squares in one row and $3+4n$ in the other.

PILE 1

PILE 2

	1	2	3	4	5	6	7	8	9	10	11	12	13
1	L	W	W	W	L	W	W	W	L	W	W	W	L
2	W	L	L	W	W	L	L	W	W	L	L	W	W
3	W	L	L	W	W	L	L	W	W	L	L	W	W
4	W	W	W	L	W	W	W	L	W	W	W	L	W
5	L	W	W	W	L	W	W	W	L	W	W	W	L
6	W	L	L	W	W	L	L	W	W	L	L	W	W
7	W	L	L	W	W	L	L	W	W	L	L	W	W
8	W	W	W	L	W	W	W	L	W	W	W	L	W
9	L	W	W	W	L	W	W	W	L	W	W	W	L
10	W	L	L	W	W	L	L	W	W	L	L	W	W
11	W	L	L	W	W	L	L	W	W	L	L	W	W
12	W	W	W	L	W	W	W	L	W	W	W	L	W
13	L	W	W	W	L	W	W	W	L	W	W	W	L

Table A.11

Lightning Source UK Ltd.
Milton Keynes UK
26 January 2011

166433UK00001B/18/P